Recognizing a danger reduces it by half.
But to recognize it, you need to know a few things.

avalanche.

Recognizing the decisive problems and danger patterns

The standard work of snow and avalanche science by Rudi Mair and Patrick Nairz

Translated from the German by Jeffrey McCabe

Tyrolia Verlag · Innsbruck Vienna

contents.

preface.

On 8 April 2009 an experienced mountain rescue squad member heads out on a backcountry skiing tour into the Stubai Alps to the Zischgeles, just under 3000 m. Nocturnal skies were clear and star-studded, in early morning the snowpack is frozen hard. It promises to be superb firn-snow joy as a reward for getting up early. The skiing descent ends in tragedy. The man dies at 10:50 am in a slab avalanche which he himself triggers at 2000 m altitude.
On 9 April 2007 – two years prior, almost to the day – six backcountry ski tourers decide at 11:20 am on an equally flawless spring day to ski down the 40° steep, east-facing gully before them. Three persons are swept away by a slab avalanche. The father of a young family is found dead by his comrades beneath two and a half meters of snow.

To jump ahead: It is no coincidence that both avalanche accidents occurred at similar times of year and of day, in similar spots, under similar conditions. On the contrary, precise analysis enables us to uncover a system behind them, like a red thread which weaves its way through all our knowledge of snow and avalanches. A system which can roughly be dubbed 'patterns' and which it is the aim of this book to elucidate through numerous examples drawn from real life experiences. The book accumulates the knowledge of 25 years of accident analysis of the Tirolean Avalanche Warning Service (LWD Tirol). The big advantage: the patterns described herein can immediately be perceived (and their lessons applied in outlying terrain) by interested skiers and freeriders, deepening practical, personal understanding of complex situations and linked conditions in snow and avalanche science.

With best wishes for accident-free experiences in wintery terrain,
Patrick Nairz and Rudi Mair

experiences & developments.

When our practical manual entitled "avalanche. Recognizing the 10 decisive danger patterns" was first published at the onset of winter 2010/11, no one dreamed that it would become a best seller of snow and avalanche research in the shortest time. The danger patterns contained therein provided a decisive impulse to epoch-making developments in the avalanche warning services, all of which had the same goal: to give users a practical, easily understandable aid to better evaluate and understand avalanche dangers.

□ Developments worldwide

A year after the introduction of our 10 decisive danger patterns, descriptions of "typical avalanche situations" or "avalanche problems" or "danger patterns" appeared in a number of avalanche bulletins, this time not only in Europe but also the U.S. (Utah and Colorado) and Canada. The background was always the same: avalanches are not coincidences. They can be causally assigned to various schemes which can be easily grasped by users. Since there are somewhat different descriptive formats from country to country, what mattered as of that point was to develop a unified concept within the avalanche warning services. Following initial attempts, the decisive impulse for a unified path came from Europe, this time from the avalanche warning experts and communications service in Slovenia. They underscored the importance of thinking in patterns, but also pointed out that avalanche warning services need to concentrate on what is essential, what is visible and what is easily graspable

Avalanche 'problems' and 'danger patterns' are a supplementary aid towards improved analysis and understanding of avalanche dangers. The object is to positively influence the on-site behavior of backcountry skiers and freeriders and to reduce avalanche accidents.

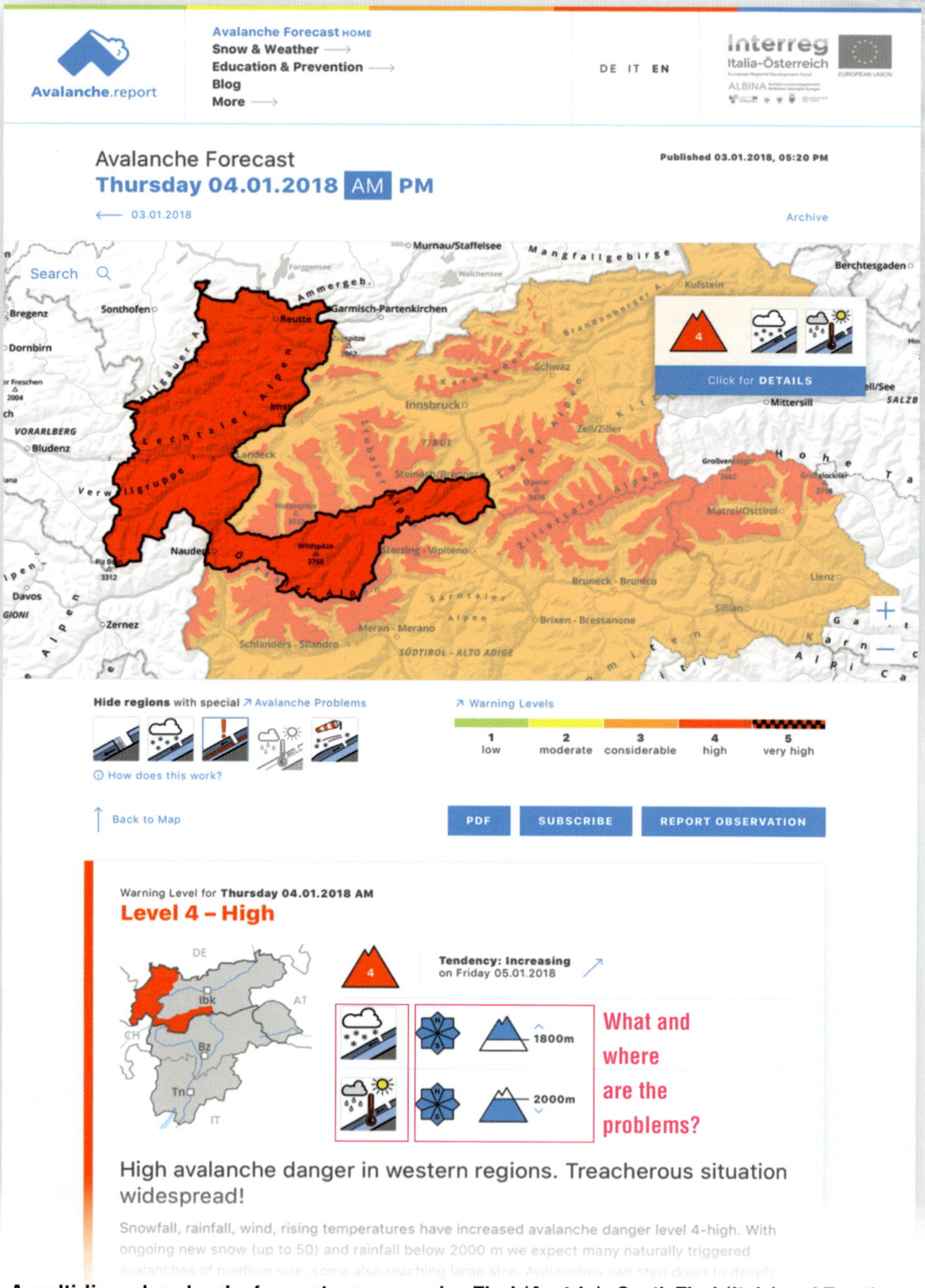

A multi-lingual avalanche forecast encompassing Tirol (Austria), South Tirol (Italy) and Trentino (Italy) will be published daily starting in winter 2018/19.

in public perception. Details might be relevant to the experts, but should not figure too prominently in a user manual. Furthermore, the method of communication should be uniform and consistent in order to avoid any confusion in the minds of users. They proposed a three-part concept for distributing information according to the following scheme:

What is the problem?
Where is the problem?
Why does the problem exist?

In 2017 the European Avalanche Warning Services officially agreed on introducing a device they dubbed the 5 'avalanche problems' – new snow, wind-drifted snow, persistent weak layers, wet snow and, optionally, gliding snow. In addition to that, each avalanche service is permitted to categorize situations with low avalanche danger as a 'favorable situation'. The 5 avalanche problems are easier for grasp by users than the 10 danger patterns; for that reason they are more prominently displayed in the avalanche bulletin. Nevertheless, the danger patterns actually increase understanding of the 5 avalanche problems, since each problem could be assigned to one or more patterns. The 10 danger patterns reveal more details and explain the processes which led to the problem (why does the problem exist?). The user thus receives more and deeper information about the prevailing avalanche dangers. That additional information, however, is more complex and presupposes a certain store

of knowledge. The overall contents of the avalanche bulletin need to be seen and grasped quickly. What is the problem? Where does the problem occur? Why does the problem exist? The last question is what gives rise to a danger pattern. In the time which has passed since this device was first introduced, many avalanche warning services use the 5 avalanche problems as an initial orientation, while the 10 danger patterns bring about deeper understanding of the situation.

□ Structure of this book

This sixth, completely revised and supplemented edition of our book treats both subjects. It is subdivided into three main text sections: Principles; The 5 Avalanche Problems; The 10 Decisive Danger Patterns.

In the principles, which are elucidated in the introductory chapter of prior editions of this book, we concentrate on the big points which are essential for a clear understanding of the overall contents. What has been added to them are some interesting statistical reinforcing data regarding distribution of danger patterns and their categorization in avalanche accidents which have occurred over the last five years, an important impulse for avalanche education.

The central section of the book addresses the 5 avalanche problems and the 10 decisive danger patterns (dp). The individual danger patterns are described in tried-and-tested fashion, each with a theoretical and a practical part, the latter with a brief accident analysis newly added.

problems & danger patterns.

Avalanche problems and danger patterns have one thing in common: they point to typical, visible and recurring situations of danger. The distinction lies in the focal point. Avalanche problems permit a rough overview of potential sources of danger, e.g. new snow. Whereas danger patterns permit penetrating insight into the causes of the situation, e. g. problem brought about by too much new snow atop a weak layer. Danger patterns thus describe possible scenarios or processes (over time) which in turn give rise to the avalanche problem (in a certain place). The aim is clear: avalanche problems and danger patterns help users to recognize dangerous situations more swiftly so that human behavior can be suitably adapted. And avalanche accidents thereby avoided.

principles.

Why did we write this book?

Every winter we stand at the rim of snow craters where shortly prior people have lost their lives. Our task is the analysis of the accident, often carried out with the Alpine Police. These situations are a spur to human reflection, especially when it is obvious that people fall into the selfsame traps again and again. Over the course of decades, it has become clear to us that even during highly varied winters, nearly identical avalanche situations still take shape. The basic structure comes from a combination of the given snowpack layering with the weather conditions added to it, together setting off an avalanche. Experienced winter ski tourers manage, often quite unconsciously, to 'call up' the appropriate avalanche situation from their rich storehouse of knowledge and memory and to adapt their conduct accordingly. But most backcountry skiers simply do not have the time to immerse themselves in the nuances of snow layering, weather and avalanches to such a degree. Thus, they see only a wintery landscape without recognizing the hidden dangers which lurk beneath the surface. What helps in that situation is the regionally published avalanche bulletin which points out the danger spots. Through this supportive aid, not only the geographical distribution of avalanche prone locations in outlying terrain can be delineated; also the degree of additional loading which the snowpack can bear before fracturing and releasing an avalanche; or whether avalanches can trigger naturally without any external stimulus at all. The system has proven itself to be of value, particularly in connection with applying risk strategies, e.g. reduction method, stop or go, etc.

spot of burial

Danger level		Snowpack stability
5 very high		The snowpack is poorly bonded and largely unstable in genera
4 high		The snowpack is poorly bonded on most steep slopes.
3 considerable		The snowpack is moderately to poorly bonded on many steep slopes.
2 moderate		The snowpack is only moderately well bonded on some steep slopes, otherwise well bonded in general.
1 low		The snowpack is well bonded and stable in general.

Nevertheless, we know that skiers and freeriders, despite instruction, explanation and elucidation, still have problems interpreting the danger levels. This is partly because one danger level does not denote merely one clearly defined situation, but a number of different ones, e.g. danger level 3 can describe a situation close to danger level 2 (moderate) or close to danger level 4 (high).

In the interim, the daily avalanche bulletin has gained new momentum. It now presents itself in the best obtainable design of quick-and-easy understanding. What is most important comes first. Maps provide a quick overview. New media with cutting-edge data, icons and images add further support and enhance our view that a picture is worth more than a thousand words. We are clearly on the right path. In our day-to-day practice of snow and avalanche science there is further potential to transmit our knowledge more simply, more rapidly and more effectively to winter sports lovers. That's why we elucidate the avalanche problems which lie behind dangerous situations in terms of clearly defined danger patterns (dp) in detail. With this system, we have been able to take great strides whenever it was necessary to analyze and portray complex and critical situations. The system has been refined continually over the years and at the same time its communicability and clarity simplified for the general public. In avalanche bulletins nowadays, attention is called first to the danger level; then to the specific avalanche problems which prevail; then to the causal danger pattern. This is done in order to deepen and intensify a snapshot grasp of the big picture: What (is dangerous) – Where (is it dangerous) – Why (is it dangerous)?

Likelihood of triggering	Tips, suggestions for all those outside secured ski runs
Numerous very large-sized and often extreme natural avalanches can be expected, even in moderately steep terrain.	Backcountry ski tours not possible in general.
Triggering is likely, even from low additional loads, on many steep slopes. In some cases, numerous large-sized and often very large-sized natural avalanches can be expected.	Backcountry ski tours require high ability to assess dangers on-site, possibilities severely reduced.
Triggering is possible, even from low additional loads, particularly on the indicated steep slopes. In some cases large-sized, in isolated cases very large-sized natural avalanches are possible.	Backcountry ski tours require experience in danger assessment on-site. Reduced possibilities.
Triggering is possible, primarily from high additional loads, particularly on the indicated steep slopes. Very large-sized natural avalanches are unlikely.	Favorable conditions as long as local dangers taken into consideration.
Triggering is generally possible only from high additional loads in isolated areas of very steep, extreme terrain. Only small and medium-sized natural avalanches are possible.	Generally safe conditions for backcountry ski tours.

avalanche-endangered terrain described in detail in avalanche bulletin e.g. altitude, aspect, type of terrain etc

moderately steep terrain slopes shallower than 30 degrees

steep terrain slopes steeper than 30 degrees

very steep, extreme terrain adverse slope angle (more than 40 degrees), terrain profile, proximity to ridge, smoothness of underlying ground surface

low additional load individual skier / snowboarder, riding softly, not falling; snowshoer; group with good spacing (minimum 10m) keeping distances

high additional load two or more skiers / snowboarders etc. without good spacing (or without intervals); snowmachine; explosives; single hiker/climber

For other special terms and definitions please consult glossary **www.avalanches.org**

The 10 decisive danger patterns (dp) still form the focal point of the sixth edition of this book, of course. We describe them in calendar order. We use the same simple scheme: first, a short description of the dp; then, the story of an accident and a detailed analysis. The avalanche accident was the direct outcome of that specific danger pattern. Then follows a precise definition. Background conditions of the pattern are brought to light. Important factors are pointed out to help recognize the given dp. Additional current examples help to entrench understanding of the system, conclude the presentation of a dp.

□ Who is this book for?

This book is written most of all for active backcountry skiers and freeriders. It is written for amateurs and for experts. The latter, we hope, will gain something from the accident analysis and discover new aspects here and there. The amateur will find new knowledge in the fundamental information about snow and avalanche science. For better comprehensibility, we begin with the fundamentals. Following that, the structure we have chosen permits us greater freedom of explanation. Whoever prefers to proceed systematically is advised to read the book in the sequence in which it was written. That way, the journey leads from the 5 avalanche problems through the world of the 10 decisive danger patterns from the start of winter to its end.

However, if a reader wishes to first accumulate some knowledge for a springtime tour in outlying terrain, he would be well advised to first look at the "wet-snow" or the appropriate "springtime situation" (dp 10). Each problem, each pattern, can be examined all by itself.

□ What are patterns?

Patterns are encountered all the time in everyday life, not just in avalanche science. As regards the patterns introduced here in the form of avalanche problems and patterns, they are situations and/or processes which recur and which can be sub-divided and categorized. The aim is to uncover clearly defined, frequently recurring, visible situations of danger.

This is not a brand new approach. It was already used in the initial development of avalanche fore-casting models about forty years ago. At the time, the focus was on historic avalanches which caused damages, for which a variety of weather parameters for that period (including overall snow depths) were available. Current weather parameters were compared with the historic data. Then sit-uations were sought in which the specifics most closely resembled those of the historic event. Then one examined whether avalanches were documented on those days and concluded that the current day might well have similar avalanche activity to the historic day. This system proved its initial value in detecting visible dangers.

dp. definition

Patterns are encountered all the time in everyday life, not just in avalanche science. As regards the patterns introduced here in the form of avalanche problems and patterns, they are situations and/or processes which recur and which can be sub-divided and categorized. The aim is to uncover clearly defined, visually apparent and frequently recurring signs of danger.

To cite one example, a significant connection between avalanches which had been observerd and the combination of intensive snowfall, strong wind impact and rapid temperature change was recognized and mathematically defined. But in complex situations the system tended to break down. That was because the most important parameter, namely, the current layering of the snowpack, had not been taken into consideration.
The definition of avalanche problems is based on possible sources of danger, e.g. new snow, wind-drifted snow, etc. Whereas with danger patterns, the focal point is on the unfolding processes which brought about the current snowpack layering and to the anticipated changes which will occur as a result of forecast weather (the current snowpack layering is seen as the immediate result of the preceding weather). Another way of differentiating them is that avalanche problems occur only in a certain place, whereas danger patterns occur from a sequence of events. In other words, the first is an occurrence in space; the second, in time.
If you look at airline risk management, certain parallels come to mind. There, too, complex, non-perceptible processes are taking place which (can) lead to dangerous situations. In our case, the platform is a physically not precisely definable snowpack; airlines use highly technological instruments in which isolated individual systems can be far more precisely defined, compared to snowpack analysis. Yet in both fields, there are interactions which take place within the systems. Furthermore, there are external (environmental) influences which also exert an impact. And in the final overall picture, the human being stands at the center of the system. Human behavior can influence the system

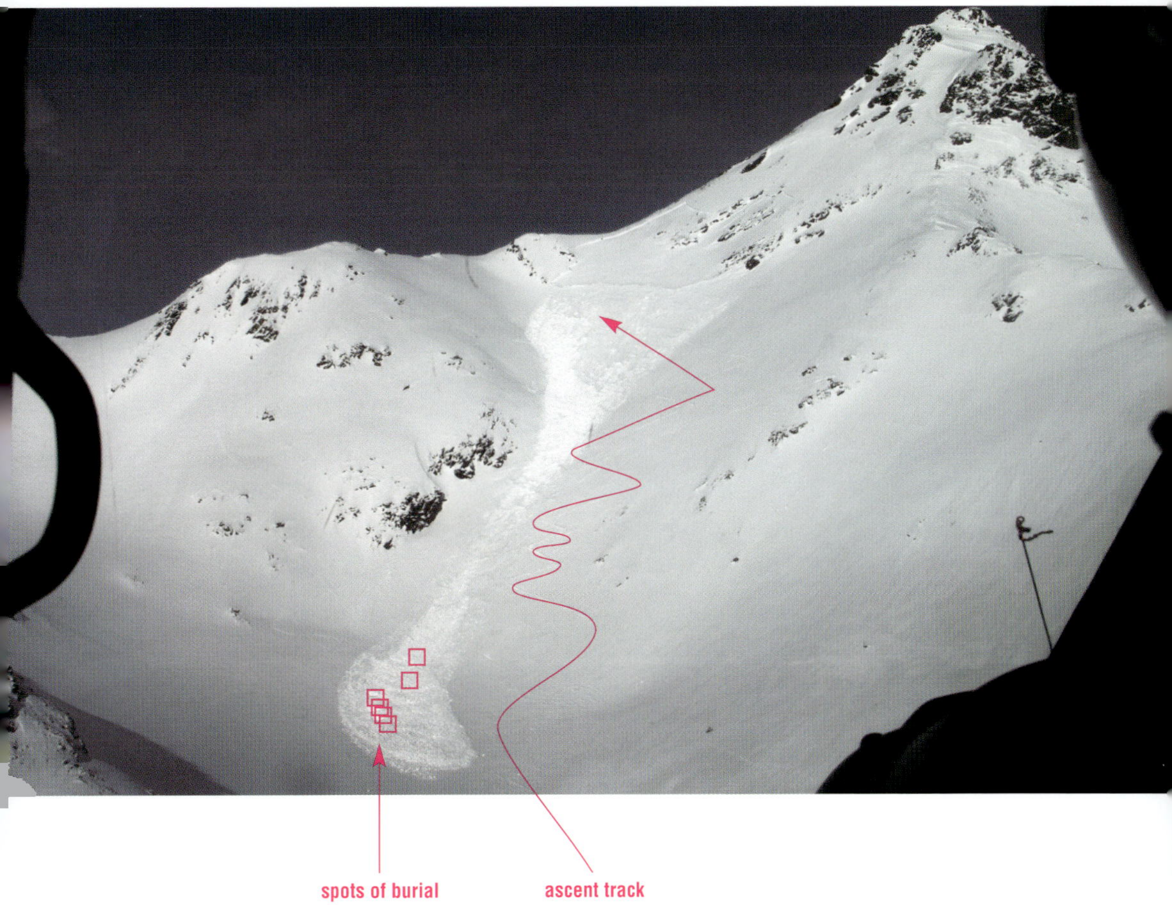

Analysis of avalanche accidents uncovers the danger pattern. Thus, analysis is an integral part of this book:

Where Sulzkogel / northern Stubai Alps / 2950 m / SE-slope / 40°

Who 15 persons involved / 4 injuries / 3 fatalities

When 22. 2. 2005, appx. 2:30 pm

Avalanche slab (dry-snow) / L 250 m / W 45 m / fracture 0.3–0.7 m / buried 1 m / 30 min.

Regional danger level 3 (considerable)

Aval. bulletin headline considerable danger regionally in high alpine

Avalanche problem new snow / persistent weak layers

dp. frequency

The individual danger patterns occur at highly varying frequencies. The dp "snowfall after long period of cold" (dp.5) numbers among the patterns which cause the highest number of avalanches relative to its frequency. In absolute numbers, however, most accidents occur in the danger pattern "loose snow and wind" (dp.6).

both positively and negatively. Amidst all these highly complex interactions the trick lies in defining the limits of the danger situations as clearly and precisely as possible. Subsequently, they have to be depicted, i.e. described or pictured as simply and strikingly as possible. For only when a danger is recognized, and correctly recognized, can countermeasures be taken.

To cite an example from the airline sector, the third attempt to land an airplane, after the first two attempts have failed, is considered to be an especially threatening danger pattern. In avalanche science, "snowfall after long period of cold" (dp.5) is a danger pattern in which (proportionate to its frequency of occurrence) the highest number of avalanche accidents occur.

Data from many accidents during the last two and a half years have helped us to develop what originally was a vague idea into a loosely-arranged sequence of danger patterns and ultimately into a cohesive system. The 10 decisive danger patterns cover at least 95% of all dangerous situations which arise during the course of a given winter. What characterizes these patterns above all else is their recurrence not only within one winter season, but over the course of highly diverse winter seasons. To make the danger patterns easy to recognize, we have concentrated on their labels, reworked them repeatedly, and finally decided on these descriptive names:

An overview of the 10 decisive avalanche danger patterns

dp.1 deep persistent weak layer
dp.2 gliding snow
dp.3 rain-on-snow
dp.4 cold on warm / warm on cold
dp.5 snowfall after long period of cold
dp.6 loose snow and wind
dp.7 shallow snow next to deep snow
dp.8 buried surface hoar
dp.9 buried graupel
dp.10 springtime situation

The 5 avalanche problems are drawn from the five potential sources of danger in use since 2014/15:

- new snow
- wind-drifted snow
- persistent weak layers
- wet snow
- gliding snow

Distribution of general danger levels between winters 1993/94 and 2017/18

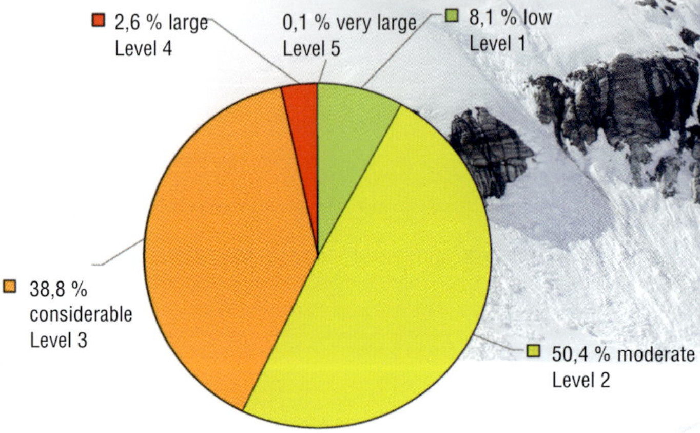

Classification options

The core of a pattern lies in its classification. It seems logical to proceed from large to small, that is, from the overview to the details. A given winter season consists, fortunately, not only of dangerous situations. There are also long periods of favorable conditions. Our danger level statistics (recorded after the five standard European danger levels were originally introduced) corroborate this. According to these statistics, in the 25 winters between 1993/94 and 2017/18, Danger level 1 prevailed an average 8.1% of the time; Danger level 2 for 50.4% of the time. Danger level 3 prevailed 38.8% of the time; and Danger level 4 prevailed for 2.6% of the time. (Danger level 5 occurred only for a few days.) Another classification possibility is found in terms of snowpack stability and danger patterns. This book sets its focus on recognizing avalanche problems and danger patterns. In the following subdivisions divided in time and space, we concentrate quite consciously on danger patterns, since the term 'avalanche problems' is too broad.

Chronological breakdown

The 10 decisive danger patterns occur within a variety of time spans. For that reason it is logical to arrange things in chronological sequence when pinpointing the relevant dp.

Chronological occurrence of individual danger patterns over the course of a winter.

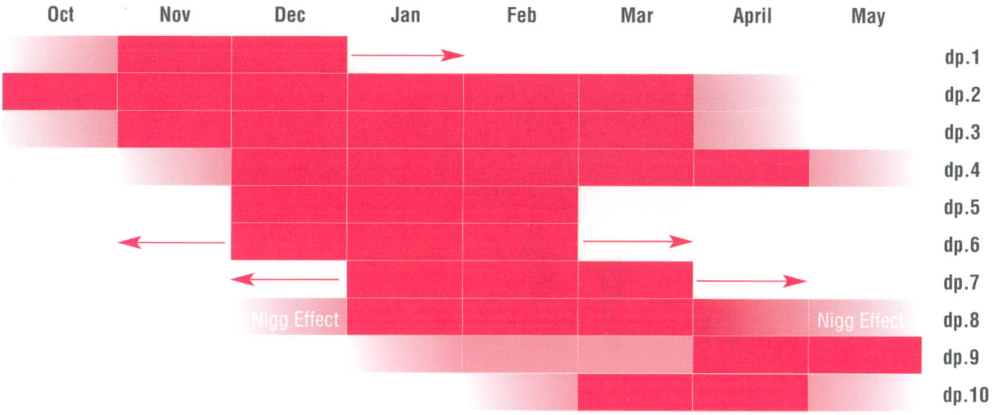

Danger pattern 1 (deep persistent weak layer) can occur in late autumn in high alpine regions; springtime situations ordinarily occur as of the end of February. Since these are natural processes, there are no hard and fast calendar boundaries in danger patterns, but rather, free-flowing transitional phases, also exceptional circumstances occasionally. Such, for example, is the case when at the height of winter, or even in early winter, springtime conditions set in briefly. It was a phenomenon observed on 10 January 2005. At the time the snow depths were less than average and the snowpack was quite unfavorably layered over widespread areas. The unusual temperatures on that day had a very negative effect on the snowpack. Small and medium-sized naturally triggered avalanches were the result. A similar situation occurred during the 2009/10 winter at the beginning of February. The prevailing conditions were identical: a pronounced weak layer was buried by new, poorly bonded snow. Warm, springlike weather weakened the snowpack to such an extent that avalanches triggered naturally. Conversely, classic high-winter situations can also occur in springtime.

When considering aspects which relate to time rather than territory, it is conspicious that certain danger patterns lead to a bundling of avalanche events. In the section “Statistics and Consequences for Everyday Practice” details about this can be consulted.

In chronological classification it is not just a matter of when a particular dp occurs during the course of a given winter, but how long the situation lasts.

For example, dp.3 (rain-on-snow) occurs relatively seldom and usually for only a short period. Whereas dp.2 (gliding snow) can persist on steep, grass-covered slopes for the entire winter as a latent threat. For dp.5 (snowfall after long period of cold) a classic period of low temperatures is the prerequisite, as in high winter. The weak layers which form in such a period can remain prone to triggering for weeks. Whereas dp.9 (buried graupel) generally occurs in late winter. Furthermore, the graupel layer which serves as a weak layer for avalanches tends to bond rather swiftly with the layer of snow deposited on top of it, due to the ordinarily higher temperatures.

□ Territorial Classification

Each dp occurs most often at certain altitudes and in certain aspects, in some cases even in certain regions: For example, dp.3 (rain-on-snow) occurs equally in all regions, but is more frequent at low and intermediate than at high altitudes. Whereas dp.4 (cold on warm / warm on cold) most readily occurs where there are great temperature disparities inside the snowpack. This can be the case when a warm front with rainfall is followed by a cold front with snowfall. Such conditions are also possible follow-ing a warm period without precipitation, when the snowpack is superficially moist at least in sunny terrain. If temperatures then drop and snowfall follows, one can usually observe great temperature disparities at the borderline between old and new snow, which is essential to the formation of weak layers.

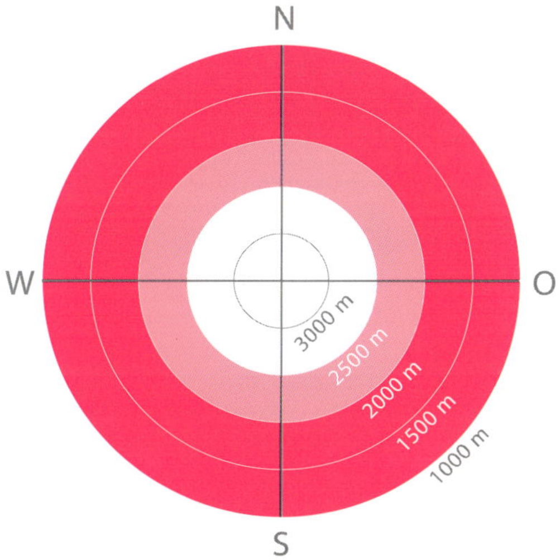

Territorial distribution of dp.3 rain-on-snow

- occurrence probable
- occurrence possible
- occurrence unlikely

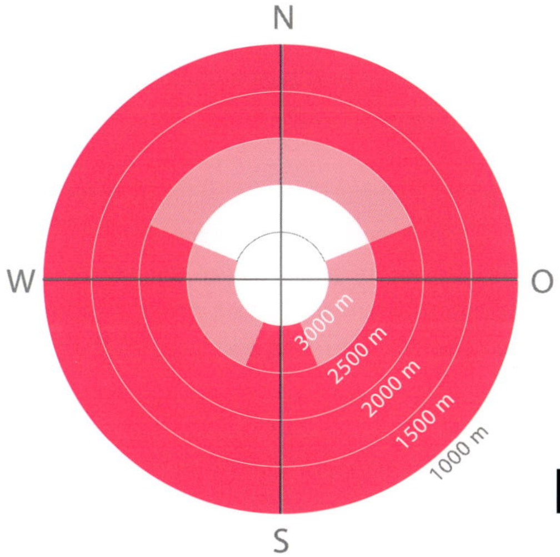

Territorial distribution of dp.4 cold on warm / warm on cold

- occurrence probable
- occurrence possible
- occurrence unlikely

dp. how far a dp extends territorially

In the case of dp.9 (buried graupel) the decisive grains of graupel accumulate in concavities in the terrain, typically in bowls and below rocks. Far less frequently in this dp is there a pronounced, area-wide weak layer which can lead to large-sized avalanches.

Altitude is what generates the set of conditions for dp.4: conditions for dp.4 are enhanced at low and intermediate altitudes; and in sunny, higher altitude zones.

The territorial extent of a dp is also fascinating. One example was the occurrence of dp.4 during early winter 2007/08. Following a foehn-induced phase of prounounced warmth during which the uppermost layers of the snowpack became thoroughly wet up to approximately 2600 m, snowfall followed bringing 20 cm of new snow accompanied by swiftly dropping temperatures. In widespread areas of Tirol, a treacherous weak layer formed in the wake of this occurrence. Three avalanche fatalities were the consequence.

Apart from the spread of various danger patterns at certain altitudes, in certain aspects and in certain regions, there is another territorial aspect to consider, namely, the spatial extent of a weak layer. For example, in the case of dp.9 (buried graupel) the decisive grains of graupel normally accumulate in concavities in the terrain, typically in bowls and below rocks. Far less often in this dp is there a pronounced, area-wide weak layer which can lead to large-sized avalanches. Quite the opposite is the case with dp.8 (buried surface hoar). This can extend over vast, cohesive areas, particularly following periods of cold. It occurs far more seldom in small areas, e.g. on the banks of streams or as the result of the Nigg Effect, explained in dp.8.

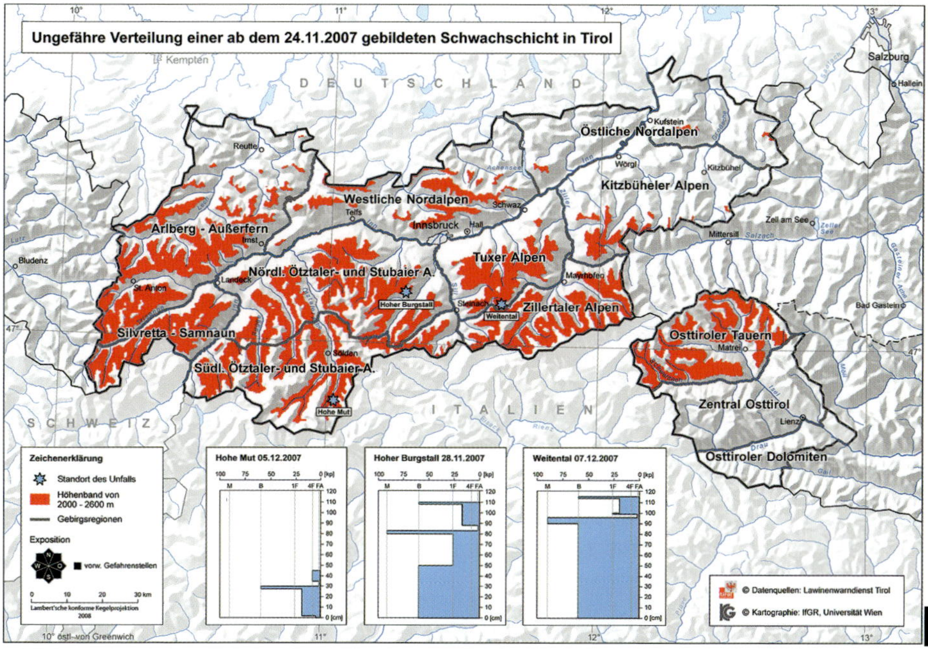

Ungefähre Verteilung einer ab dem 24.11.2007 gebildeten Schwachschicht in Tirol
D E U T S C H L A N D
S C H W E I Z
I T A L I E N
Östliche Nordalpen
Westliche Nordalpen
Kitzbüheler Alpen
Arlberg - Außerfern
Nördl. Ötztaler- und Stubaier A.
Tuxer Alpen
Zillertaler Alpen
Silvretta - Samnaun
Südl. Ötztaler- und Stubaier A.
Osttiroler Tauern
Zentral Osttirol
Osttiroler Dolomiten
Kempten
Salzburg
Hallein
Kufstein
Wörgl
Kitzbühel
Reutte
Telfs
Schwaz
Innsbruck
Hall
Imst
Landeck
St. Anton
Bludenz
Zell am See
Mittersill
Mayrhofen
Steinach
Sölden
Matrei
Lienz
Bad Gastein
Hoher Burgstall
Weitental
Hohe Mut
10° östl. von Greenwich
11°
12°
13°
47°
Zeichenerklärung
Standort des Unfalls
Höhenband von 2000 - 2600 m
Gebirgsregionen
Exposition
N O S W
vorw. Gefahrenstellen
0 10 20 30 km
Lambert'sche konforme Kegelprojektion 2008
Hohe Mut 05.12.2007
Hoher Burgstall 28.11.2007
Weitental 07.12.2007
100 75 50 25 0 [kp]
M B 1F 4F FA
120 110 100 90 80 70 60 50 40 30 20 10 0 [cm]
© Datenquellen: Lawinenwarndienst Tirol
© Kartographie: IfGR, Universität Wien

dp. overlapping

During wintertime, distinct avalanche problems and danger patterns sometimes overlap, although one is usually dominant.

☐ Overlapping danger patterns

Avalanche situations can usually be allocated to one avalanche problem or one decisive danger pattern. This is simplest at the beginning of winter and in its final phase. During the course of winter, however, various avalanche problems and danger patterns sometimes overlap, even if as a rule one of them is dominant. To cite a short example from winter 2009/10: on 29 December we were in outlying terrain below the Kellerjoch in the Tux Alps in order to investigate an avalanche accident. We cut and examined snow profiles in several places. It became clear that the uppermost layer of snow was heavily wind-impacted. Beneath that was a thin, loose layer of fresh snow. Everything pointed to dp.6 (loose snow and wind), the new fallen snow being the weak layer. When we dug deeper into the snowpack, however, we uncovered a thin layer of faceted snow crystals just beneath an equally thin melt-freeze crust. This weak layer had formed during a shift from cold to warm (dp.4) after 25 December. Deeper still, we found a very loose layer which formed during a lengthy period of cold (dp.5) before 20 December. Two backcountry skiers triggered the fresh layer of snow- drift during the ascent. Through additional loading on the gliding snow masses, the more deeply embedded weak layers subsequently broke away.

This example reveals a challenge in applying danger patterns. The conditions which led to dp.5 lay further back in time, but proved equally relevant for the overall snow layering. Thus, the question

Beneath a beautiful blanket of snow, dangerous weak layers often lurk.

Weak layers form through a variety of processes which can be understood and recognized. For the avalanche in the background, dp.6 was primarily responsible.

sometimes arises of how to point out these things in the most transparent way. One elegant solution would be that of our colleagues from across the Atlantic, namely, to subdivide the weak layers into three basic sub-groups, independent of their formation process. They speak of "deep persistent weak layers" and "persistent weak layers" and "non-persistent weak layers" in this connection. These are persistent weak layers close to the ground; or weak layers inside the old snowpack; or new snow as a weak layer (see chapter dp.6). The last of these three can be triggered only for a brief interval; the first two are triggerable over longer periods of time.

If we try to imagine the situation pictorially, the various weak layers which formed through quite different processes and which can be described by danger patterns tend to work their way down to more deeply embedded levels of the snowpack where they can persist as more-or-less dangerous layers which are prone to triggering. The process of formation and expansion is then of interest only to professionals.

Avalanche problems are an excellent approach when providing an overview. For example, this method permits us to point out a persistent weak layer problem in an all-encompassing way. The danger pattern(s) responsible for it can also be pointed out, but need not necessarily be described. As always, detailed information can be obtained in the text of the avalanche bulletin or via new media. Overlapping occurs not only in danger patterns, but also in avalanche problems. Many possible combinations

Where Kellerjoch / Tux Alps / 2200 m / SE-slope / 38°

Who 2 persons involved

When 28. 12. 2009, 12:30 pm

Avalanche slab (dry-snow) / L 200 m / W 70 m / fracture 0.3 m / burial 10 min.

Regional danger level 3 (considerable)

Aval. bulletin headline Considerable avalanche danger above 2000 m

The avalanches near the Zischgeles referred to in the Preface.

are imaginable. For example, if lots of new snow has been deposited on top of an unstable old snowpack surface, one can assume a combination of a new snow and persistent weak layer problem. New snow and wind, on the other hand, often results in a combined new snow / wind-drifted snow problem, etc.In addition, avalanche problems can also overlap. For example, if surface hoar is buried by new snow (dp.8) or a lengthy period of cold comes to an end with snowfall (dp.5), we assume a new snow problem. Following several days without precipitation, a persistent weak layer problem forms as a result of the the trigger-sensitive weak layer over a long period.

Long years of experience have shown us that the same avalanche problems and danger patterns can overlap both territorially and in time. In other words, avalanches trigger repeatedly under the self-same pre-conditions at similar times in the same places. This is the case regularly for the avalanche problem gliding snow and dp.2 (gliding snow).

An impressive example of such overlapping in time and territory is found in the example which was referred to in the Preface near the Zischgeles. In two gullies adjacent to each other, two avalanche fatalities occurred due to the wet snow problem, danger pattern 10 (springtime situation) separated by exactly two years (almost to the day). The prerequisites as regards snow layering and weather conditions, as well as the additional loading of the skiers involved, were identical. ▌

Where Zischgeles – Sattelloch / northern Stubai Alps / 2320 m / E-slope / 35°
Who 3 persons involved / 1 fatality
When 8. 4. 2009, 10:50 am
Avalanche slab (wet snow) / L 400 m / W 40 m / fracture 0.3–1 m / burial 0.5 m / 1 day
Regional danger level 1 (low) -> 3 (considerable)
Aval. bulletin headline Perfect touring conditions in the morning – Increasing avalanche danger during the day
Avalanche problem wet snow

Where Zischgeles – Köllrinne / northern Stubai Alps / 2280 m / E-Hang / 35°
Who 6 persons involved / 1 fatality
When 9. 4. 2007, 11:20 am
Avalanche slab avalanche (wet) / L 250 m / W 80 m / fracture 0.4 m / burial 2.3 m / 1 hour
Regional danger level 1 (low) -> 3 (considerable)
Aval. bulletin headline Favorable touring conditions widespread, daytime rise in danger
Avalanche problem wet snow

statistics & consequences for everyday.

As we all know, every winter is different. This is evident not only in the highly varying snow depths and avalanche fatalities. The fluctuating prevalence of danger patterns is also a telltale sign. A glance at the dp distribution tells you everything you need to know about a given winter.

For example, the winter of 2011/12 in the northern regions of Tirol was a "grainy" winter. Following several dangerous phases with vast amounts of new snow came a period of very favorable conditions. A short span in mid-February stood out in which nearly all the avalanches of the season took place. The cause lay in snowfall accompanied by wind impact following a long period of cold (dp.5). The winter of heavy snowfall was also characterized by pronounced danger of glide-snow avalanches. Relative to the potential risks, relatively few avalanches occurred. Whereas the winters 2014/15 to 2016/17 unfolded quite differently, characterized by a massive persistent weak layer problem. That means that over a lengthy period of time, a hidden danger lurks inside the snowpack which can be dealt with only through enormous restraint on the part of skiers. What also stands out is the disproportionately high percentage of casualties during a persistent weak layer problem, compared to the more frequently occurring wind-drifted snow problem. Eight years have passed since the danger patterns were first introduced. This period of observation has supplied us with such a wealth of information that in the interim valuable interpretations of statistics can be presented. Interpreting the danger patterns and assigning them to avalanches reveal interesting trends which are useful in protecting against future avalanche accidents. All in all, the following stands out:

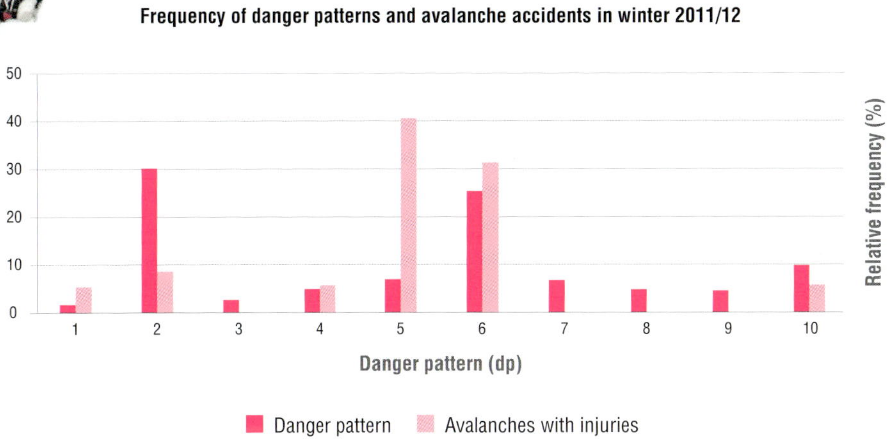

dp.5 (snowfall after long period of cold) led to nearly half of all avalanches with casualties in winter 2011/12

What is striking about the distribution of danger patterns during winters 2014/15 to 2016/17 is the persistent weak layer problem which prevailed because of dp. 1 (deep persistent weak layer) with above average avalanche frequency.

□ Prevalence of danger patterns and allocation to accidents in winters 2010/11 - 2017/18 in Tirol

dp.1 (deep persistent weak layer): numerous accidents and injuries, many involving experienced, trained people.

dp.2 (gliding snow): often cited, but accident occurrences were low.

dp.3 (rain-on-snow): Backcountry skiers and freeriders are rarely surprised by rain.

dp.4 (cold on warm / warm on cold): Significant weak layers are often found near the upper surface in strictly limited altitude zones and aspects. The location of the weak layer presupposes heightened proneness to triggering by skiers, on the one hand; and rapid deterioration of the layer by atmospheric conditions on the other. Numerous accidents with casualties.

dp.5 (snowfall after long period of cold): corresponds to our experience in winter 2011/12; in a short period of time, frequent avalanche accidents.

dp.6 (loose snow and wind): This is the most frequent danger pattern; it is also the cause of most avalanche accidents. This should be a focal point of avalanche training, particularly since the danger pattern is so apparent and in most cases is easily recognized with some experience.

dp.7 (shallow snow next to deep snow): This dp can be linked to an existing persistent weak layer problem, both over a small surface or a wide area, which is included in other danger patterns. For that reason, this danger pattern is statistically less significant.

Frequency of danger patterns and avalanche accidents in winter 2014/15 to 2016/17

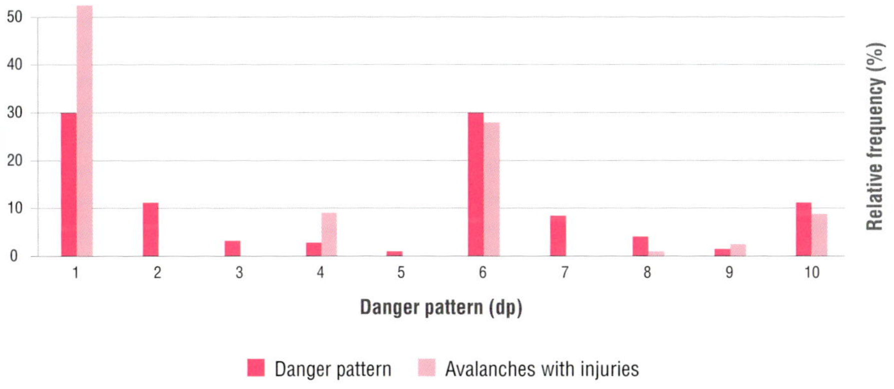

dp.6 recommendation

When dp.6 occurs – loose snow transported at low temperatures – the wind-loaded slopes should be avoided whenever possible. Ordinarily only one day of patience is required before conditions improve significantly.

❙ **dp.8** (buried surface hoar): We expected a prevalence similar to dp.5 but were surprised. One possible explanation for the low numbers of accidents: we use this dp also for the small-surface Nigg Effect.

❙ **dp.9** (buried graupel): As expected, this dp seldom causes accidents, if at all, then in springtime.

❙ **dp.10** (springtime situation): Relatively high danger potential, but casualties were not above average.

What remains is the recognition that in danger patterns where weak layers form (e.g. faceted snow crystals, depth hoar or surface hoar) accidents are far more frequent. This is particularly the case for dp.1, dp.5 and dp.8, and often for dp.4.

At this juncture, we appeal to all backcountry skier and freeriders: when dp.6 occurs – wind transporting loose snow at low temperatures: avoid drifted slopes whenever possible. Ordinarly, only one day of patience is required before conditions improve significantly. ❙

Frequency of danger patterns and avalanche accidents in winter 2010/11 to 2017/18

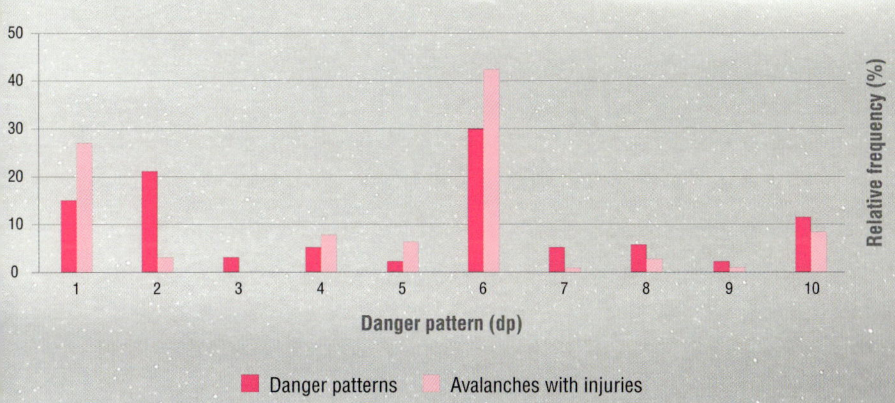

the 5 avalanche problems. (europe)

Avalanche problems, like danger patterns, point to typical, recurring and generally visible danger situations. The difference lies in the focal point. Whereas avalanche problems provide a wide-angle overview of potential sources of danger, danger patterns look deeper into the material and pinpoint the reasons the problems came about. The objective is clear: dangerous situations can be recognized more quickly with the help of avalanche problems and danger patterns, you can adjust your behavior and avoid avalanche accidents.

☐ Avalanche problem: New snow

Behavior: Wait until the new snow has stabilized and/or ski defensively.
New snow does not automatically mean that avalanche danger increases. As so often in avalanche science, that depends on several factors. What is decisive is that the problem is always directly linked to the new fallen snow; and that there needs to be a critical mass of new snow in order for the problem to arise.

Experience has taught us that snowfall frequently occurs together with wind impact. Thus, we refer to a combined new snow / wind-drifted snow problem. However, when there is snowfall without wind, either loose-snow avalanches in rocky terrain are the danger, or slab avalanches if the new snow bonds with the snowpack due to higher temperatures.

new snow problem

What?	**Characteristics**		This avalanche problem results from current or the most recent snowfall. The amount of additional loading from new snow on the existing snowpack is the crucial factor of the new snow problem. How critical the loading is depends on various factors such as air temperature or characteristics of the old snow surface.
	Expected avalanche types		▌Dry-snow slab avalanches ▌Dry loose-snow avalanches ▌Natural and human-triggered avalanches
Where?	**Spatial distribution**		Generally widely present and often in all aspects.
	Location of weak layer in the snowpack		Usually at or close to the interface to the old snow surface, sometimes in the new snow layers and sometimes slightly below in the old snowpack.
Why?	**Trigger mechanisms**	**Dry-snow slab avalanches**	Additional load due to snowfall on existing weak layers (old snow surface or below) or newly created weak layers (within the new snow)
		Dry loose-snow avalanches	Lack of cohesion between the new snow particles
When?	**Duration**		Typically during snowfall and up to a few days after.
How to manage?	**Identification of the problem in the field**		The new snow problem is fairly easy to recognize since it affects most of the terrain. Focus on critical amounts of new snow and recent avalanche activity.
	Advice	**Dry-snow slab avalanches**	Wait until the snowpack has stabilized and the weak layer has adapted to the new load and gained strength.
		Dry loose-snow avalanches	The danger of falling is more important than the danger of burial. Focus on consequences in steep terrain.

Possible danger patterns are dp.1 / dp.4 / dp.5 / dp.8 / dp.9

In case of extremely heavy snowfall, on the other hand, the weight of the new snow alone can become so great that weak layers break away and avalanches grow to large size. Special situations in which new snow is also an immediate factor are discussed in the chapters on dp.4 and dp.9. The "advantage" of the new snow problem is that it is usually easy to recognize.

☐ Avalanche problem: Wind-drifted snow

Behavior: Heed signs of wind, avoid snowdrift accumulations.
Wind and snow transported by wind are the major elements of the wind-drifted snow problem. Snow can be transported during snowfall (resulting in a combined new snow / wind-drifted snow problem) or after the snowfall has come to an end.

What is essential for understanding the problem are the weak layers inside the snowpack on top of which the snowdrifts are deposited. It makes a huge difference what kinds of snow crystals the weak layers are composed of, which in turn has an effect on how long they will persist. Particularly long-lived are depth hoar, faceted snow crystals and surface hoar. The wind-drifted snow problem is often accompanied by weak layers expanding inside the snowpack. This is particularly the case when loose powder snow gets buried by wind-drifted snow at low temperatures.

Wind-drifted snow problem

What?	**Characteristics**	This avalanche problem results from wind-drifted snow. Snow can be transported by wind with or without concurrent snowfall.
	Expected avalanche types	▍Dry-snow slab avalanches ▍Natural and human-triggered avalanches
Where?	**Spatial distribution**	Highly variable but typically on leeward slopes in gullies, bowls, near distinct changes in slope angle, behind ridgelines or other wind-sheltered locations. More frequent above treeline.
	Location of weak layer in the snowpack	Usually at or close to the interface to the old snow surface or within the slab layer due to variations in wind speed during storm cycle, but sometimes slightly deeper down in the old snowpack.
Why?	**Trigger mechanisms**	Wind-drifted snow is an additional load on a weak layer and builds a slab that is particularly prone to intensify fracture propagation.
When?	**Duration**	The wind-drifted snow problem can develop very quickly and typically lasts during the snowdrift event, up to a few days at most, depending on snowpack evolution.
How to manage?	**Identification of the problem in the field**	If not concealed by new snow the wind-drifted snow problem can be recognized with training and good visibility. Focus on wind signs and locate deposits. Typical clues: snowdrift deposits, recent avalanche activity and sometimes shooting cracks or whumpfs. Moreover, it is often hard to determine the age of wind signs and wind signs do not necessarily imply an avalanche problem (e.g., in absence of a weak layer).
	Advice	Avoid snowdrift deposits in steep terrain.

Possible danger patterns are dp.1 / dp.4 / dp.5 / dp.6 / dp.8

In such a situation slab avalanches can easily be triggered at the borderline between the loose powder snow, which is the weak layer, and the wind-drifted snow. For further details, see the chapter on dp.6. If you can read the indicators of wind, it is relatively easy to circumvent wind-drifted snow, unless the drifts have been covered by fresh snowfall. Most often skiers, following a period of transported snow, lack the patience to wait until the situation loses its danger.

□ Avalanche problem: Persistent weak layers

Behavior: Select a prudent route and ski defensively. Be particularly cautious where the snow is shallow and in transition zones from shallow to deep snow.
The existence of weak layers inside the snowpack is the prerequisite for this avalanche problem. Such layers are comprised by the snow which has not been subjected to the influence of precipitation or wind or melting for several days. In regions or periods where the snow is shallow, the persistent weak layer problem is more prevalent than when (and where) the snow is plentiful. It can occur in all aspects. Most often the problem is observed on shady slopes.

Slab avalanches are most prevalent are in places where the snow is shallow or in transition zones from shallow to deep snow, triggered by large additional loading. Such avalanches also occur in flat terrain.

Persistent weak layers problem

What?	Characteristics	This avalanche problem results from the presence of one or more persistent weak layers in the old snowpack. These weak layers typically include faceted crystals, depth hoar or surface hoar
	Expected avalanche types	▌ Dry-snow slab avalanches ▌ Mostly human triggered avalanches; natural avalanches are rare, mainly in combination with other avalanche problems
Where?	Spatial distribution	This avalanche problem can be widespread or isolated. It can exist in all aspects, but is found more frequently on shady, wind sheltered slopes.
	Location of weak layer in the snowpack	Anywhere in the old snowpack, often deep inside the snowpack. When deeply buried, however, triggering becomes less likely.
Why?	Trigger mechanisms	Avalanche releases when loading exceeds the strength of the weak layer.
When?	Duration	Weak layers can persist for weeks to months; possibly most of the winter season.
How to manage?	Identification of the problem in the field	Persistent weak layers are very challenging to recognize. Signs of instability such as whumpfs are typical but not essential. Stability tests can be helpful to detect the persistent weak layers. Information on snowpack history is critical and reference to the published avalanche report is important. Fracture propagation over long distances is common; remote triggering is possible.
	Advice	Select route prudently and avoid large steep slopes. Focus on the weather and snow cover processes in the area. Be extra cautious in areas with a thin snowpack and in transitions from shallow to deep snow. This problem is a major cause of recreational avalanche fatalities.

Possible danger patterns are dp.1 / dp.4 / dp.5 / dp.7 / dp.8

Fracture propagation is possible over vast distances. This is how large-sized avalanches are born. It is a problem which is difficult to assess, usually persists for a lengthy period of time, as confirmed by statistics: when experienced people die in avalanches, it is usually during a persistent weak layer problem.

What helps most in this situation is restraint on the part of skiers. And in addition to that, for trained people, snowpack analysis. Incidentally, after several days without precipitation a new snow problem develops into a persistent weak layer problem on the snowpack surface. In springtime, as the snowpack becomes wetter and wetter, one often observes a combined persistent weak layer and wet snow problem. Further details can be found in the chapters on dp.1, dp.4, dp.5, dp.7 and dp.8.

□ Avalanche problem: Wet snow

Behavior: Good timing and a prudent route selection are decisive.

This avalanche problem arises through the snowpack becoming ever more moist, sometimes thoroughly wet, thereby forfeiting its firmness. This can occur through rain, through higher air temperatures, intensive solar radiation, high air moisture or warm winds, or through a combination of these factors.

What must be examined with a particularly critical eye is the first water seepage into a weak layer (causing rapid increase in danger during winters when snow is shallow, see example on the

wet snow problem

What?	Characteristics	This avalanche problem results from a weakening of the snowpack due to the presence of liquid water. Water infiltrates the snowpack due to snowmelt or rain.
	Expected avalanche types	▌ Wet-snow slab avalanches ▌ Wet loose-snow avalanches ▌ Mainly natural avalanches
Where?	Spatial distribution	When melting due to solar radiation is the main cause, distribution of the problem depends mostly on aspect. The altitude depends mainly on air temperature and humidity. All aspects are affected in the event of rain on snow.
	Location of weak layer in the snowpack	Anywhere in the snowpack
Why?	Trigger mechanisms	**Wet-snow slab avalanches** ▌ Weakening and failure of pre-existing weak layers in the snowpack or release at layer interfaces where water accumulates ▌ Rain is also an additional load on weak layers **Wet loose-snow avalanches** ▌ Loss of cohesion between snow crystals
When?	Duration	▌ Hours to days ▌ Rapid loss of stability possible ▌ Especially critical as water infiltrates deeper down for the first time after the snowpack has warmed up to 0 °C. ▌ Natural avalanches can become more likely in the course of the day, depending on aspect (unless rain is dominating factor).
How to manage?	Identification of the problem in the field	The wet snow problem is usually easy to recognize. Onset of rain, snowballing, pin wheeling and small wet slabs or wet loose-snow avalanches are often precursors of natural wet-snow slab avalanche activity. Deep penetration is another sign of increased wetting.
	Advice	If the wet snow surface freezes into a crust during a cold night with clear sky, conditions are usually favorable in the morning. After warm overcast nights the problem often exists already in early morning. Ordinarily, rain on fresh snow creates this problem almost immediately. Good timing and trip planning are important. Focus on avalanche runout zones.

Possible danger patterns are dp.3 / dp.10

Griesskopf in chapter dp.10); massive water storage atop a hardened crust; and transition zones from snowfall to rain.

Wet-snow avalanches, regardless whether wet-snow slab avalanches, loose-snow avalanches or glide-snow avalanches, contain immense danger potential. A wet snow problem is easily visible. Good route selection and timing are important, also because the danger of a thoroughly wet snowpack can increase quite rapidly. For details, see the chapters on dp.3 and dp.10.

□ Avalanche problem: Gliding snow

Behavior: avoid all zones near glide cracks!

This avalanche problem threatens the gliding away of entire masses of the snowpack across steep, smooth surfaces. Glide-snow avalanches frequently "announce" imminent release by glide cracks in the snowpack surface. The point in time of their release is impossible to predict; in autumn they frequently trigger following large amounts of fresh snowfall or in springtime when the snowpack becomes thoroughly wet for the first time. For details, see the chapter on dp.2.

gliding snow problem

What?	**Characteristics**	The entire snowpack glides over the ground, typically on smooth ground such as grassy slopes or smooth rock-zones. High glide-snow avalanche activity typically results from a thick snowpack with no or only few weak layers. Glide-snow avalanches can occur with both a cold dry snowpack and a warm wet snowpack. The release of a glide-snow avalanche is difficult to predict, although glide cracks open usually before a release.
	Expected avalanche types	▌ Glide-snow avalanches; cold, dry or 0 °C isothermal wet snowpack ▌ Almost exclusively natural avalanches. Human and artificial triggering is unlikely.
Where?	**Spatial distribution**	Primarily on smooth ground and on slopes of any aspect, but more often on south-facing slopes.
	Location of weak layer in the snowpack	Interface between the ground and overlying snowpack
Why?	**Trigger mechanisms**	Glide-snow avalanches are caused by a loss of adhesion at the snow / ground interface.
When?	**Duration**	Days to months; occasionally during entire winter season. The release can occur at any time of day or night. In the spring, glide-snow avalanches occur most often during the latter part of the day.
How to manage?	**Identification of the problem in the field**	With the presence of glide cracks the problem can often be localized; however, they do not indicate imminent avalanche release. That is nearly impossible to predict. Avalanche release without pre-existing glide cracks is also common.
	Advice	Avoid areas close to glide cracks.

Possible danger patterns are dp.2

danger pattern.1
deep persistent weak layer

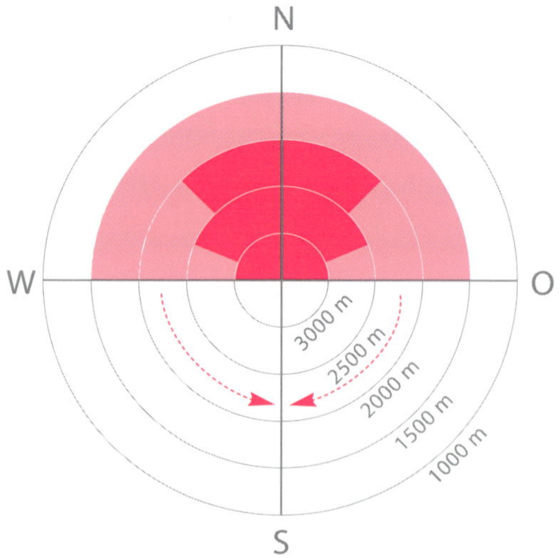

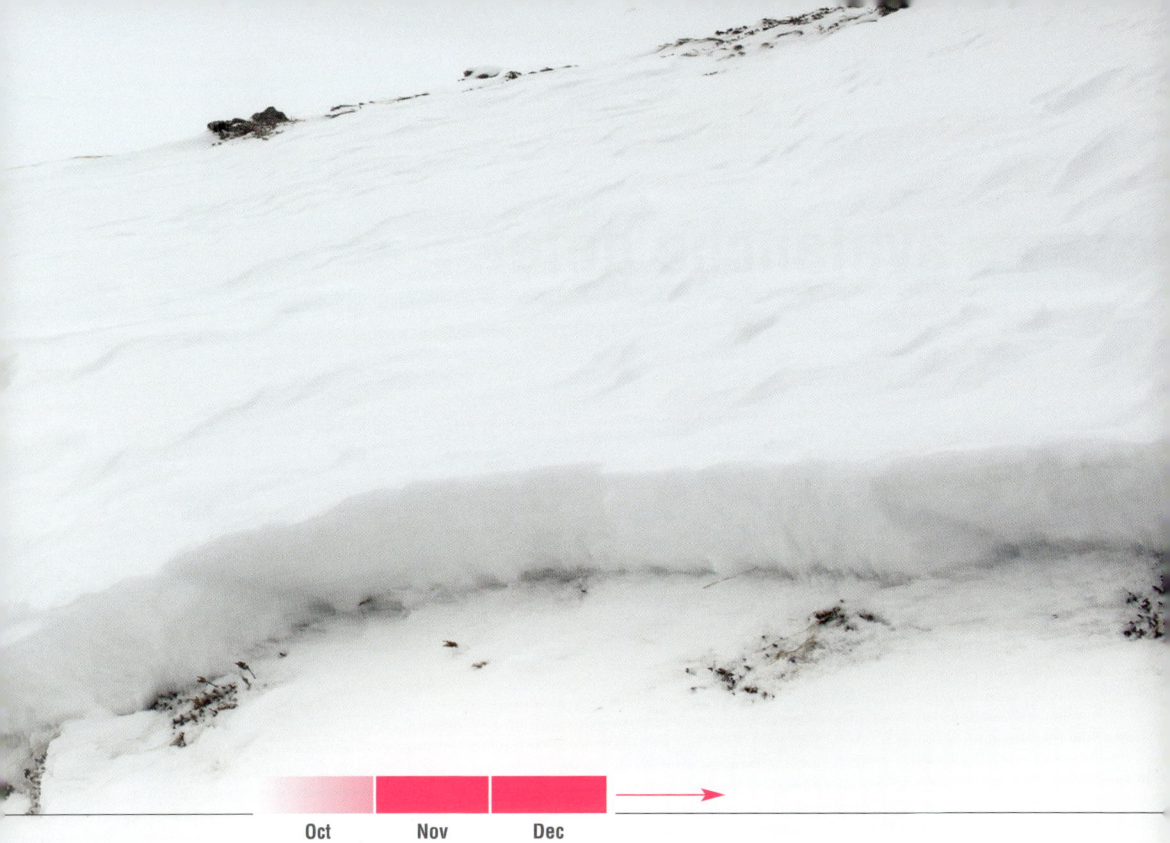

At the onset of winter, slab avalanches are often observed following the second heavy snowfall, usually at high altitudes (>2000 m) and high alpine altitudes (>3000 m) on shady, very steep slopes. This is because in autumn after the first big snowfalls, a lengthy period of stable, high-pressure weather or at least precipitation-free weather frequently follows. That is when ideal conditions prevail for the partial metamorphism of snow into loose crystals, which then form a deep persistent weak layer for subsequent snowfall to be deposited on.

During the course of early winter, this can occur in a wide range of altitudes and aspects. In general, deep persistent weak layers formed in early winter are triggered by skiers and freeriders at the beginning of the season, but it is also possible later on in winters with little snowfall. This occurs especially in spring when deep persistent weak layers are weakened further by water seepage.

avalanche geier.

In this avalanche accident on the Geier, several avalanches triggered over widespread areas. Five persons were killed. Huge media interest was awakened. The cause of the avalanche lay in a massive persistent weak layer problem, typical of the winter 2015/16.

Circumstances of the accident

Two separate Czech groups are in the Wattener Lizum. One of the groups consists of twelve persons including two freerider guides; the other consists of eight persons. The larger group cuts the track through the snow from the Lizumer Refuge on the valley floor towards the Geier. When the slope gets steep, the group turns back and, after a brief skiing descent, reassemble in a flat spot at about 2300 m altitude, where they meet the second group on its ascent.

The second group follows the same track through the snow, then continues onwards towards the Geier. Seven persons of this eight-person group easily reach a spot of flattish terrain beneath the final rise to the summit. In the interim, the twelve-person group has reattached its skins and relaunched its ascent. They are on the steep slope together with the eight-person group, maintaining proper distances between them, when suddenly a slab avalanche triggers above them. After a slight delay, five further avalanches trigger. Three of them sweep over parts of the deposit area of the primary avalanche. Ten persons are completely buried, three persons partially buried. One member of the group

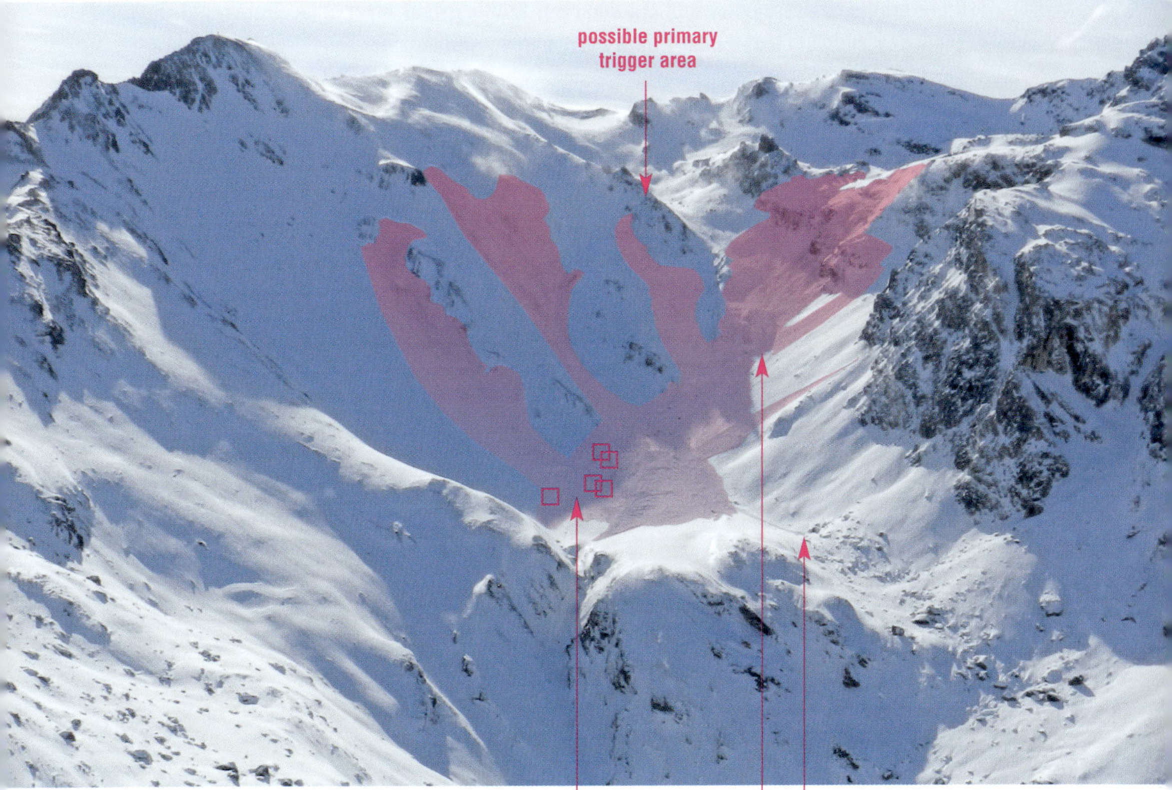

Where Geier / Tux Alps / 2620 m / NE-facing slope / 38°

Who 20 persons involved / 5 fatalities, 2 injured persons

When 6. 2. 2016, 12:10 pm

Avalanche slab avalanche (dry-snow) / L 700 m / W 150 m / fracture depth 0.6–2 m / burial 1.2 m–3.5 m / up to 3 hrs.

Regional danger level 3 (considerable)

Aval. bulletin headline Beware high altitude drifts. Persistent weak layer problem in shallow-snow zones.

Avalanche problem persistent weak layers

Snow profile near assumed spot of triggering. Primary weak layer of faceted crystals near snow saw, beneath that a melt-freeze crust with snowmelt water canal.

informs the Tirolean alarm center and a huge search action is initiated, including four helicopters. All the completely buried persons are located with a transceiver. Five of them are dead.

☐ Analysis

Weather and snowpack. Variable weather conditions prevail prior to the accident. Between 30 January and 1 February a cold front brings snowfall down to low lying areas, subsequently a warm front brings rainfall up to about 2400 m. Thereafter, temperatures drop again. Starting on 3 February, more snowfall arrives. All in all, the amount of new snow registered in the accident zone during the week prior to the accident was about 50 cm. The snowfall was accompanied by strong-velocity winds, resulting in far-reaching snowdrift accumulations being generated.

Decisive for this accident, apart from the weather conditions in the week prior to the event, was the snowpack layering. In particular, deep persistent weak layers which formed during this winter period of shallow snow and above-average temperatures. These were loose layers of metamorphosed (faceted) crystals in the vicinity of at least one thin melt-freeze crust. The weight of one skier was thus sufficient to trigger this weak layer. The persistent weak layer problem of deep persistent weak layers was quite pronounced on shady slopes of the inneralpine regions above about 2300 m, as well as on sunny slopes above about 2500 m at this juncture of the season.

loose, faceted layer
snowmelt channel

dp.1 recognizing it

The prerequisite for this danger pattern is the forming of a deep persistent weak layer. Problems occur where there is an area-wide snowpack which is then covered by bonded snow (usually snowdrifts). This is most frequently the scenario in early winter in shady, very steep terrain adjacent to ridgelines at high altitudes. That makes it quite easy to localize.

This is the first approximate turn-back spot of the group of 12. During the avalanche they are on this slope.

The accident reveals the wide-area spread of the weak layer, including pronounced fracture propagation. This chain of events necessitates, apart from the pronounced weak layer, a "perfect slab." And precisely that formed on 30 January, encouraged by the warm front, on the one hand, with rainfall extending up to 2400 m; and by the subsequent snowfall accompanied by strong winds, on the other.

As regards the primary avalanche zone, no precise cause can be pinpointed. According to accounts of participants, evidence points to the avalanche being triggered by part of the eight-person group in flat terrain, about 150 m distant from the accident avalanches. That means, in all probability, it was an astounding remote triggering.

Avalanches. During the night of 31 January in the early hours of 1 February, high avalanche danger prevails for a brief interim. Naturally triggered avalanches unleash throughout the land. Also in the Wattener Lizum, several slab avalanches are observed which release in very steep terrain, whose altitude zones and directional aspects correspond to some extent with the accident avalanches on the Geier.

The 6th of February turns out to be a day with 20 avalanche events in which people are involved. Most of them stem from deep persistent weak layers from early winter. Some of them are quite similar to the avalanches on the Geier as regards wide-area fracture propagation and remarkable remote triggerings.

Terrain. The fracture zone is on a north-facing slope between 2600 and 2700 m altitude and is very steep to extremely steep. The normal route up the Geier, the one selected by both groups, has slope gradients of about 35° in places, more often the slope angle is lower. The bowl-like terrain and the flattish slope below it have a negative effect on avalanche deposits, since they concentrate the protracted, time-delayed avalanches in the flat zone at an altitude of about 2300 m. Avalanche deposits there are estimated to be 5 meters deep.

During the avalanche emergency the rescue squad had to deal with immense avalanche waves. Numerous people took part in the rescue.

background knowledge: deep persistent weak layer.

dp.1 definition

Deep persistent weak layers tend to form through snow transformation which takes place inside the snowpack in early winter. This frequently occurs during long periods of beautiful, stable weather following the first snowfall. Weak layers frequently form near the upper crust, e.g. rain crusts, melt-freeze crusts. These surface layers are loose and usually thin. Danger arises when the weak layers are then buried by new snow.

The principle of triggering a slab: bonded snow lies on top of a weak layer near the ground, dp.1

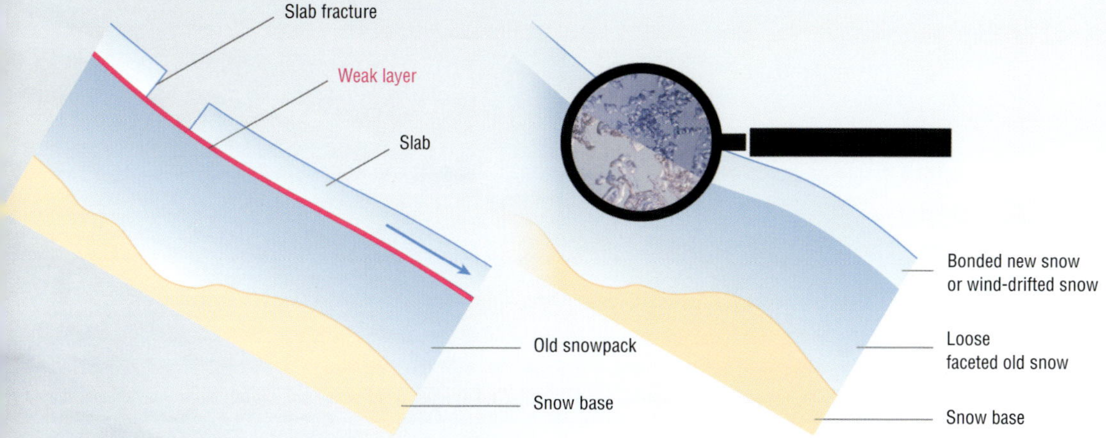

□ Slab avalanches

In order to understand why a deep persistent weak layer creates a danger pattern it is necessary to understand what a "slab" is. A slab avalanche is a type of release in which an entire mass of bonded snow ("slab") breaks away nearly explosively from the layer beneath it. The fracture is invariably sharp-edged; a wide-breadth snow mass lurches into motion in a split second. Slab avalanches are responsible for at least 95% of avalanche fatalities. Viewed in terms of time, this danger pattern can occur only after the first snowfall, since it takes a certain amount of time until a layer of metamorphosed and faceted snow crystals can form.

The more loose and unbonded a weak layer is, the less adequately it is bonded with the snow deposited on top of it when that snow itself is bonded; and the more easily it can trigger as an avalanche. The problem cited in dp.1 should never be underestimated: the pronounced and immediate increase of avalanche danger in early winter due to shallow snow depths. Even a very small slab of 20 x 20 m (=400 m^2) has a volume of 400 x 0.5 if the fracture is 50 cm deep, in other words, 200 m^3. At an average density of 200 kg/m^3 that results in (200x200) 40,000 kg, or 40 tons of snow, which corresponds to a trailer truck fully loaded with snow. A Swiss proverb quite correctly runs: "A bathtub of snow can cost your life." ▌

avalanche gaislachkogel.

Avalanche. The U.S. Ski Team is training in Sölden. During a break in the training session, six members of the team decide to descend from the Gaislachkogel over north-facing freeriding terrain, in some places steep. To begin with, they follow tracks in the snow. At an altitude of about 2400 m the group skis together into an extremely steep slope and triggers an avalanche there. Four persons manage to escape the snow masses, two young ski racers are swept along and buried beneath about 3 meters of snow in the closed descent sector of the Rettenbachtal. Since they had no emergency equipment with them, locating them and subsequently freeing them from the masses of snow takes time. Both ski racers die at the spot of the accident.

Short analysis. At the beginning of January, numerous avalanches are triggered in Tirol by backcountry skiers. The reason is the pronounced deep persistent weak layers which formed during the early part of winter. These consist of a series of thin layers of loose, faceted crystals plus rain or snowmelt crystals. They are buried by new snow which fell on New Year's Eve and was transported by winds. Stability tests show the snowpack's high proneness to triggering throughout Tirol, as well as a pronounced tendency towards fracture propagation. Settling sounds and fracture cracks forming on the surface are an everyday occurrence. Responding to this situation, the Avalanche Bulletin calls attention to the treacherous conditions for backcountry skiers and freeriders. Particularly unfavorable here is the extreme steepness of the terrain in places, as well as the fact that the slope was untracked or, at best, little frequented. ▌

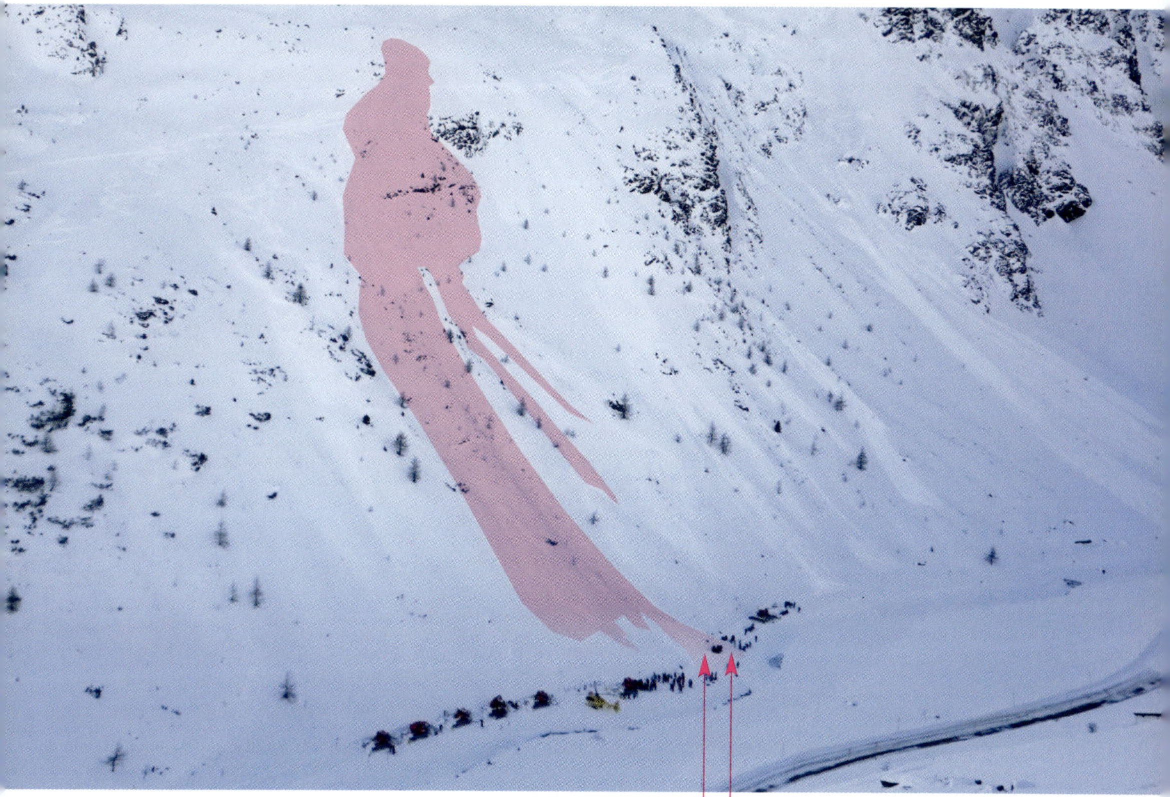

burial spots

Where Gaislachkogel / southern Ötztal Alps / 2340 m / NW-facing slope/ 42°

Who 6 persons involved / 2 fatalities

When 5. 1. 2015, 9:15 am

Avalanche slab avalanche (dry-snow) / L 350 m / W 30 m / fracture depth 0.3–0.5 m / burial ca. 3 m / appx. 1 hr.

Regional danger level 3 (considerable)

Aval. bulletin headline Treacherous avalanche situation widespread above 2200 m[2]

Avalanche problem persistent weak layers

avalanche olperer.

Avalanche. Two local mountain climbers decide to climb the Olperer (3476m). On the extremely steep north face they trigger a slab avalanche which sweeps both away and buries them. The accident is observed by another skier, who sounds the alarm and then assists the first helpers to swiftly locate one of the buried skiers. A hand sticking out of the snow is what saves the life of this person. Since the other person had no avalanche transceiver with him, the avalanche zone has to be systematically searched with snow penetration rods and avalanche dogs. One of the dogs manages to locate the person after about two and a half hours. He is buried beneath nearly 2 meters of snow and cannot be saved.

Short analysis. The slab avalanche is triggered by the mountaineers in high alpine terrain adjacent to a ridgeline on a very steep, shady slope. Snowfall in mid-September laid the cornerstone for this avalanche accident. The snow remained on the ground in this shady terrain at high altitude. Due to subsequent warm weather at the end of September, a melt-freeze crust formed on the snowpack surface. When at the beginning of October about 10-15 cm of new snow fell at very low temperatures, the temperature gradient brought about snow metamorphism and the formation of faceted, loose crystals below these crusts. In the period 9 - 12 October there was an additional 25 - 30 cm of new snow registered. Starting on 13 October, storm-strength southerly foehn winds transport this freshly fallen snow in high alpine regions. Freshly formed and hard snowdrift accumulations are generated. Stability tests indicate with near certainty that one of the descending mountaineers gave the decisive impulse to the deep persistent weak layer in extremely steep terrain. ▌

Where Olperer / Zillertal Alps / 3350 m / north-facing slope / 50°

Who 2 persons involved / 1 fatality, 1 injured person

When 16. 10. 2017, 2:10 pm

Avalanche slab avalanche (dry-snow) / L 180 m / W 40 m / fracture depth 0.35 m / burial ca. 2 m / appx. 2.5 hrs.

Regional danger level blog entry, no danger level assigned

Avalanche problem persistent weak layers / wind-drifted snow

danger pattern.2
gliding snow

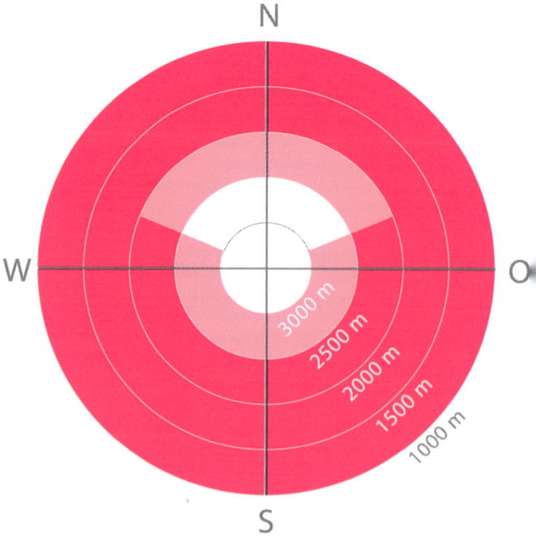

Snow glides most easily towards the valley across steep, smooth surfaces. While this process is taking place, glide cracks – easily visible, sometimes several-meters deep cracks in the snow-pack surface – are formed. Such fissures are now considered to be not the favorable signs which ancient folk knowledge once posited (such traditions are hard to overcome) but instead, utterly unfavorable indicators of potential, sometimes imminent avalanches. A glide crack points to the possibility of a glide-snow avalanche without giving any indication if or when such a glide-snow avalanche will actually release. Regarding their timing, glide-snow avalanches number among the most difficult avalanches of all to predict, since they can trigger literally at any time of day or night, even when the snow situation is at its most stable. That includes the coldest day and the warmest day of the winter. Furthermore, glide-snow avalanches are not triggered by additional loading.

avalanche hochfügen country road.

Tiny impulse. Huge effect. And some good luck. That is a brief description of the avalanche which engulfed Hochfügen's country road at the beginning of March 2012. On a very warm day, a small gliding snow mass unleashes above the road which develops into a large-sized glide-snow avalanche. One car driving uphill is buried. Lots of guardian angels must have been on duty, since only that car was buried and its driver escaped with injuries.

□ Circumstances of the accident

On 4 March just before 3:00 pm below the Kuhmesser (2264 m) in extremely steep, south-facing terrain near the ridgeline, an avalanche releases which is initially quite small. This glide-snow avalanche plummets down its path taking with it a quite moist, sometimes thoroughly wet snow-pack over wide ranging surfaces and develops into a huge wet-snow avalanche which subsequently divides into two arms. Each of these avalanches buries the public road to Hochfügen.
The orographic right arm of the avalanche buries 170 m, the left arm buries about 70 m of the road. The avalanche deposit is between 3 and 5 meters deep. The right arm buries the car of a local driver and sweeps it along with the descending snow masses. The car is totally wrecked, the driver gravely injured. A second car driving in the direction of Hochfügen with a Dutch family inside is fortunately just brushed by the snow masses. A search-and-rescue operation was launched immediately, lasted until 8:00 pm until it could be said with certainty that no other persons had been caught in the

original glide-snow avalanche

glide cracks

Where Hochfügen country road / Tux Alps / 2260 m / SE-Hang / 40°

Wer 1 person involved / 1 injured person

Wann 4. 3. 2012, 3:00 p.m.

Avalanche glide-snow avalanche (wet) / L 500 m / W 5 m (initially) / fracture 1.5 m / burial 30 min.

Regional danger level 2 (moderate)

Aval. bulletin headline Beware glide-snow avalanches!

Avalanche problem gliding snow / wet snow

Immense snow masses hamper the search.

avalanche. As a consequence of the avalanches approximately 1500 day visitors were forced to stay in Hochfügen for somewhat longer than they had planned

Analysis

Weather and snowpack. A high pressure zone which had been dictating weather conditions was starting to weaken. The day of the accident – quite similar to the days prior – was unseasonably warm. For that reason, the snowpack in the fracture zone was moist or even thoroughly wet all the way down to the ground. A melt-freeze crust which formed in the course of each night was very thin in the early morning hours and rapidly softened during the daytime. Examining the old snowpack, pronounced weak layers at this altitude and aspect were not evident. For that reason, slab avalanches were unlikely; instead, wet loose-snow and glide-snow avalanches were the threats. Their likelihood of triggering increased as the degree of snowpack wetness did.

Avalanche. The accident occurred at a time when numerous glide-snow avalanches unleashed in Tirol. For that reason, high caution was urged towards this highly unpredictable danger in the avalanche bulletins and blogs. It was strongly advised to avoid all zones below glide cracks and observe the runout zones of possible avalanches. The latter was especially urgent, since this was a winter with above average snow depths in the northern regions of the land. To backcountry skiers and freeriders,

spot where car was buried
70 / 71 gm.2

The process of the snowpack becoming wet gives rise to numerous glide-snow avalanches each year in the Innsbruck Northern Massif, as here at the beginning of March 2012.

dp.2 recognizing it

Although the time of a glide-snow avalanche release is impossible to predict, at least the place is easy to recognize. Such avalanches usually have clearly visible glide cracks for lengthy periods. In addition, observations show that over the course of several winters, glide-snow avalanches frequently trigger in the very same spots.

such warnings are surely helpful and relatively simple to heed. But safety authorities reach their limits when trying to warn of the dangers of glide-snow avalanches. The only measures which can be taken are long road closures; snow removal wherever there are glide cracks; or road protection constructions. Such drastic measures are possible only in isolated cases, e.g. on easily accessible slopes approaching ski areas. The artificial triggering devices installed in the fracture zone were ineffective against glide-snow avalanches. The only possible alternative in this situation would have been artificial triggerings of loose-snow avalanches through explosives in the fracture zone. And such a release might well have swept away the entire snowpack with it. Even then, it requires a great deal of experience to pinpoint the right moment for such an action. This becomes evident when one looks at our efforts to do precisely that two days prior, which were without success.

Terrain. The fracture zone is an extremely steep, southeast-facing slope and thus, intensively exposed to solar radiation. It can be assumed that the snowpack at the time of release was thoroughly wet and that the avalanche released abruptly. The abiding lesson is that the first time the snowpack becomes thoroughly wet in a season, numerous naturally triggered avalanches are unleashed, including glide-snow avalanches. This insight needs to be taken into consideration by public safety authorities. In the end, it must be conceded that glide-snow avalanches, along with ice avalanches, number among the most unpredictable and hard-to-assess types of avalanches. A 100% degree of safety will never exist in future, despite knowledge and highest possible safety measures. ▌

background knowledge: gliding snow.

dp.2 definition

Gliding snow is the movement of the entire snowpack parallel to the slope. Such movements are much influenced by the degree of rawness of the ground. The smoother the ground surface, the more often glide-snow avalanches will occur in steep terrain. Glide-snow avalanches are observed most often on steep, grass-covered slopes or in smooth rocky terrain.

Creeping movements at an avalanche fracture edge, following very heavy snowfall in extremely steep terrain.

Creeping and gliding movements

Fundamentally, every snowpack in sloped terrain moves to a small extent towards the valley as the simple consequence of gravity. This occurs in the course of the snowpack settling process without the formation of any glide cracks. The snowpack remains anchored to the ground; only the upper layers of snow actually move. This movement is called "snow creeping." The steeper the terrain, the deeper the snowpack, the closer to the uppermost surface the creeping occurs, the more intensive it is and the bigger the tensions unleashed inside the snow cover. In the case of "gliding snow" the snowpack itself also moves slowly across the ground. The entire snowpack all the way down to the ground can fracture and release. Glide cracks form on the surface. Snowmelt can cause water to seep to the base of the snow cover down through the glide cracks. As a result, glide-snow avalanches can be unleashed. A glide crack can be present all winter long without a glide-snow avalanche ever triggering. Conversely, avalanches occasionally release within a short period (within one hour) of a glide crack appearing on the surface.

What is decisive is the surface between the ground and the snow

The most important criterium for a possible avalanche is the structure of the layer between the snow and the ground. Free flowing water or a thin film of water at this point enhances the gliding

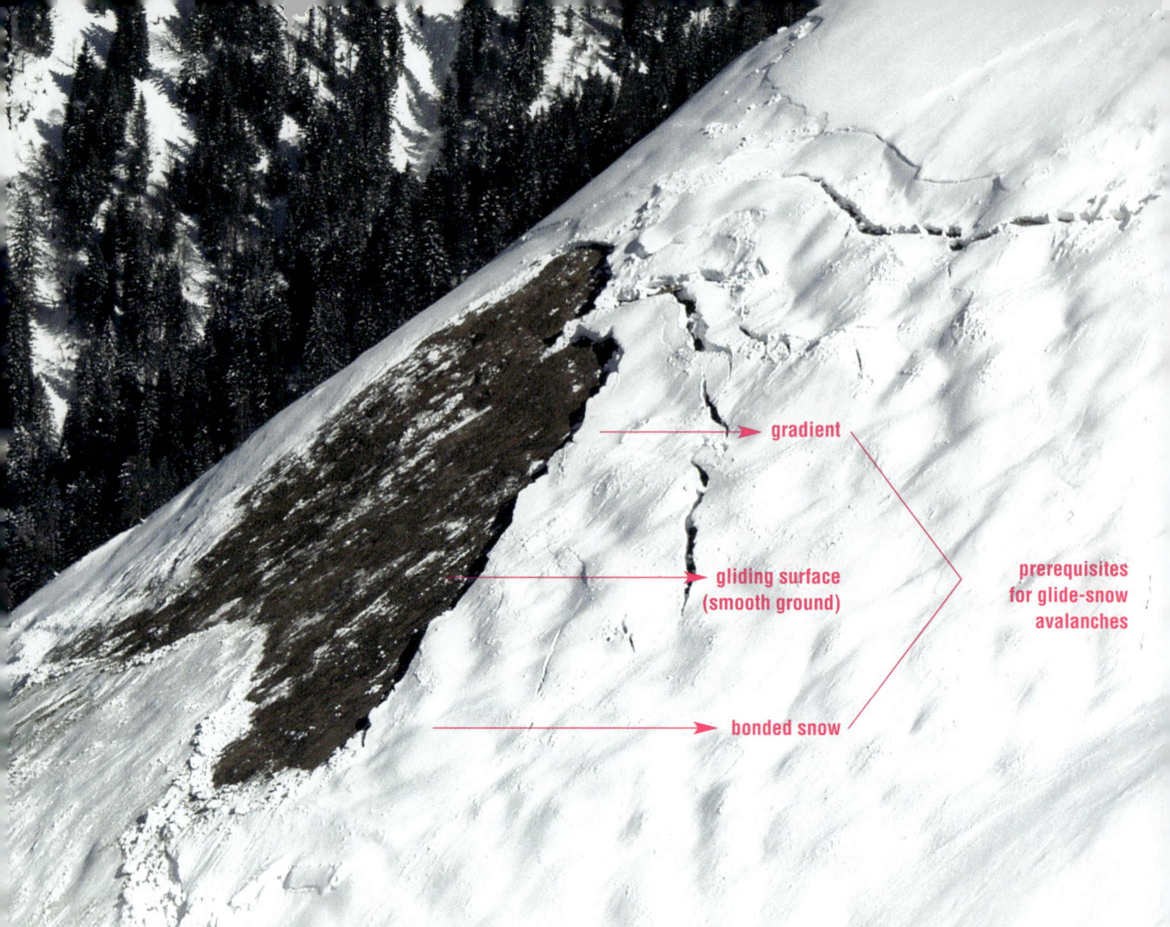

movement and, as a consequence, the likelihood of glide-snow avalanches triggering. Thus, rainfall and significantly rising temperatures increase the probability of (glide-snow) avalanches. Early, massive blanketing with snow acts to further that likelihood, since the ground temperature is then relatively warm, the water film more pronounced.

The likelihood of an avalanche triggering increases with each continuous gliding movement. The reason: such gliding motions normally increase significantly just prior to every release.

□ Trigger mechanism

If you compare a slab avalanche with a glide-snow avalanche, they have certain traits in common: the gradient of the slope needs to be right: typically, slopes which are steeper than 30°. For both types of avalanches you need a gliding surface; and for both types, the snow deposited atop the gliding masses must be bonded. That is the only way that their force can be transferred over vast surfaces. A sharp-edged fracture is also typical of both types. The biggest difference: the bed surface of a slab avalanche is snow; fracture propagation is unleashed via a weak layer. Whereas with glide-snow avalanches, the snow mass glides directly across highly variable ground surfaces, or across a firm base. There exists a wet smear-layer, but no weak layer of snow through which fracture propagation would be possible. ▌

dp.2 practical tip

In planning your route, try to avoid zones below glide cracks. Maintain a respectable distance in the runout zones of this type of avalanche.

possible avalanche path

avalanche innsbrucker nordkette arzler scharte.

Avalanche. The Arzler Scharte above Innsbruck entices more and more backcountry skiers each year, regardless whether to the powder snow in high winter or to the corn snow in springtime. One of the more enticing days for backcountry skiing tours was 1 March 2012. Following a night of clear skies, a descent launched early enough in the day guaranteed fine corn snow on the steep south-facing slope. For that reason, a group starts out early on their tour towards Arzler Scharte. What surprises them is the fresh, area-wide deposit of a huge glide-snow avalanche which washed over a large part of the route. The avalanche had triggered naturally during the previous night.

Short analysis. This is a region where snowfall is ususually heavy, during a winter which will go down in history as a "glide-snow avalanche winter." On steep, grass-covered slopes there are glide cracks over widespread areas, indicators of gliding movements of the snowpack. If we examine this tour with a magnifying glass, we also see (easier from a distance than during a tour) laterally displaced glide cracks extending way above the ascent route. Admittedly, in planning a tour up to Arzler Scharte, not even the locals place glide-snow avalanche danger in the foreground, since the ground on this surface is roughened up by widespread rock gravel. Glide-snow avalanches are simply not possible across such ground. Also surprising was the time the avalanche released: during a night of clear, star-studded skies. ▌

glide-snow avalanche

Where Innsbruck Northern Massif – Arzler Scharte / Northern Alps, western sector / 2100 m / SW-slope / 35°

Who no persons involved

When 1. 3. 2012, 1:00 am

Avalanche glide-snow avalanche (moist) / L 600 m / W 70 m / Fracture 2.5 m

Regional danger level 2 (moderate)

Aval. bulletin headline Daytime rise in avalanche danger – wet snow avalanches in very steep, wet terrain

Avalanche problem gliding snow

avalanches arlberg.

Avalanche. At the beginning of 2012, numerous skiers and freeriders were caught in glide-snow avalanches. The following description is intended to encapsulate the events of that period in which persons suffered only minor injuries. Freeriding terrain is constantly in demand on the Arlberg. Avalanche dangers do not cause great headaches to the boarders and skiers. They descend the slopes without worries. Some of them are not aware that traps are lurking in the form of glide cracks. Thus, it can happen that when visibility is poor, open glide cracks are not perceived in time; or treacherous snow bridges form which can break through when any weight hits them. Several times, rescues from crevasses have taken place in the Arlberg region, and we are not referring to glaciated terrain.

Short analysis. The Arlberg-Ausserfern region is well known to get more snowfall than most regions in Tirol. Corresponding to the relatively high proportion of grass-covered slopes, glide-snow avalanches occur quite frequently. The snow depths in winter 2011/12 were far above average and many glide-snow avalanches were the consequence. It was no surprise that some slopes which were riddled with glide cracks looked like glaciers when viewed from the opposite side of the valley. Equally unsurprising is that, given the high frequency of freeriders in Arlberg terrain, some of them break through the snow bridges into the glide cracks or are surprised by open glide cracks. A fall is then unavoidable down to the very bottom of the crack, in other words, all the way down to the grass-covered ground which is hopefully buried with snow. That winter, the depths of the falls were enormous: up to nine meters!

Where Arlberg / Arlberg – Außerfern / 2100 m / SE-slope / 35°

Who several persons involved

When January 2012

Avalanche glide-snow avalanche (dry) / fracture up to 9 m

Avalanche problem gliding snow

danger pattern.3

rain-on-snow

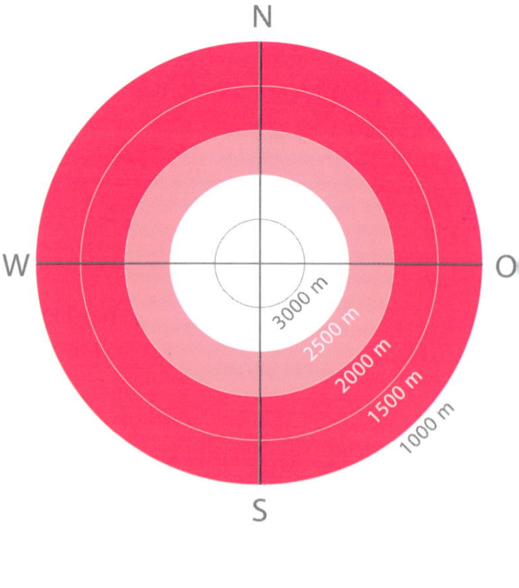

Rain-on-snow is considered the classic alarm signal in snow and avalanche science because, on the one hand, its additional weight burdens the entire snowpack; and on the other, because the firmness of the snowpack is immediately lost. For those reasons, avalanches are predestined. Rain can occur at any time during the winter. The big advantage is that no danger sign could be simpler to recognize.

avalanche niedertal.

In the statistics, loose-snow avalanches are seldom seen to cause damages. When they do, they are usually wet, loose-snow avalanches. The loss of firmness caused by rain plays a major role in the following avalanche accident in which a British soldier suffered grave injuries.

☐ Circumstances of the accident

A 15-man group of British soldiers was climbing along the Niedertal from Vent towards the Martin Busch refuge. The route along the access road leads past the bottom of extremely steep, east-facing slopes. There, a wet loose-snow avalanche triggered naturally which expanded as it plummeted downwards and caught one of the ascending soldiers in its path. He was swept into the stream bed below where he finally came to a stop on the snowpack surface with grave craniocerebral injuries. He was resuscitated and flown to Innsbruck Hospital.

☐ Analysis

Weather and snowpack. In this case, the capricious month of April lived up to its reputation. Until the 12th day of the month, it was full of variable conditions until the day of the accident. The days prior were full of mercurial shifts: heavy snowfall, cloud, sunshine and wind. On the day of the accident it was warm to start with. There was rainfall in the early part of the day which extended up to 2300 m.

Spot where avalanche caught him

Where Niedertal / Southern Ötztal Alps / 2330 m / E-slope / 45°

Who 15 persons involved / 1 person injured

When 12. 4. 2013, 1:00 pm

Avalanche loose wet-snow / L 300 m / W 1 m

Regional danger level 3 (considerable)

Aval. bulletin headline Considerable avalanche danger below about 2000 m

Avalanche problem wet snow

Trail to Martin Busch refuge with avalanche deposits and place of fall.

The problem in the snowpack is found on the surface, where at least at low and intermediate altitudes the recently fallen new snow is moist or wet. Elsewhere, a mostly stable old snowpack dominated at the time of the accident. What was decisive was the rain, falling on top of an already moistened snowpack. This led to an immense loss of snowpack firmness.

Avalanches. At this juncture of the season, the major threat for skiers and freeriders stems from wet loose-snow avalanches, in isolated cases from glide-snow avalanches on steep, grass-covered slopes. The loose-snow avalanches are a relatively short-lived threat, yet highly visible, downright obvious. The following could be read in that day's avalanche bulletin: "The major danger today stems from wet loose-snow avalanches wherever the snowpack is wet at least on the upper surface, in other words, at altitudes below about 2300 m."

Terrain. Loose-snow avalanches occur only in extremely steep terrain with gradients of 40° and more. As a rule of thumb in practice, such terrain is usually riddled with rocks. Upon closer examination, this held true in the avalanche starting zone of this accident. It is also recognizable at once that there are several avalanche starting zones which threaten this ascent route to the Martin Busch refuge. ▍

dp.3 recognizing it

Rain-on-snow is the simplest danger pattern to recognize. During rainfall in moderately cold snowpacks (-5 to 0°C) it has to be assumed that the danger level sharply increases due to a big loss of snowpack firmness. One exception is when the snowpack is very cold, in which case the rainwater is immediately bonded through freezing. The loss of firmness takes place only after the snowpack has been "warmed up" by the rain and nears the melting point (0°C).

background knowledge: rain-on-snow.

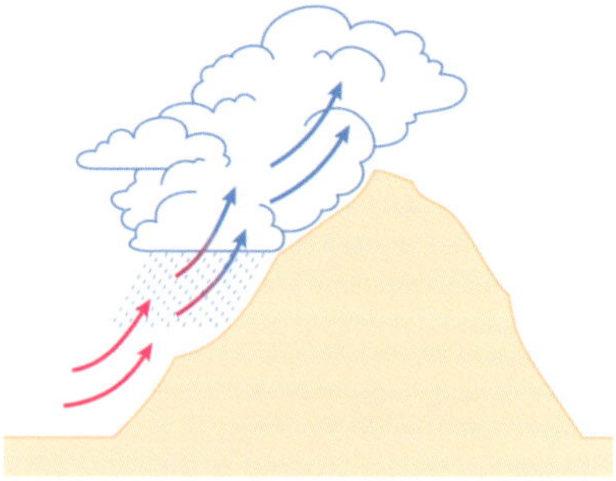

Barrier cloud precipitation

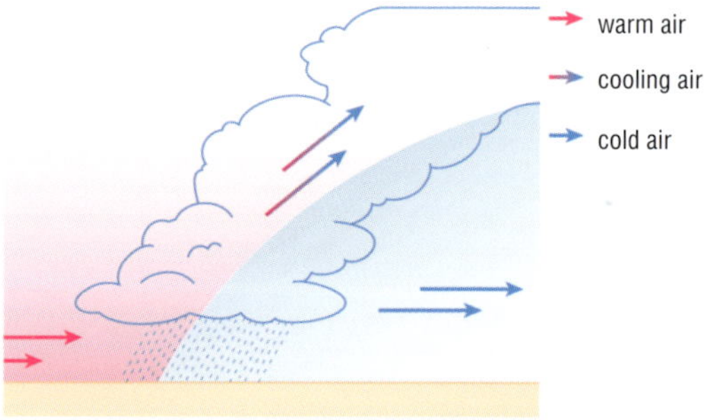

Front precipitation – warm front

How it comes about

Rain is born inside clouds. Clouds, consisting either of water drops or fine ice crystals depending on their altitude and temperature, are formed out of moist air masses. When of water drops it is called condensation; when of ice crystals, it is called deposition. In both cases, tiny elements comprising a so-called condensation nucleus, e.g. dust particles, are required. The small water drops which form keep on growing either through water vapor, other droplets or ice crystals. In other words, they get heavier and sooner or later can no longer be held by the clouds as winds penetrate them. Obeying the laws of gravity, the water falls to the ground. The typical diameter of such raindrops is between 0.5 and 3 mm. One clue in the weather report of whether it will rain or snow is supplied by the designation of the zero-degree level and/or the snowfall level. The snowfall level always lies lower down than the zero-degree level.

The uplifting force of moist air masses is invariably what is responsible for initiating the precipitation to begin with. When air masses are elevated, they cool down, condensation (or deposition) takes place and precipitation is the consequence. For the most part, this uplift is brought about by three different processes: by congestion of elevated air masses, which is typical of the mountains; by fronts; and by convection.

dp.3 definition

Rain is the most prevalent form of precipitation.

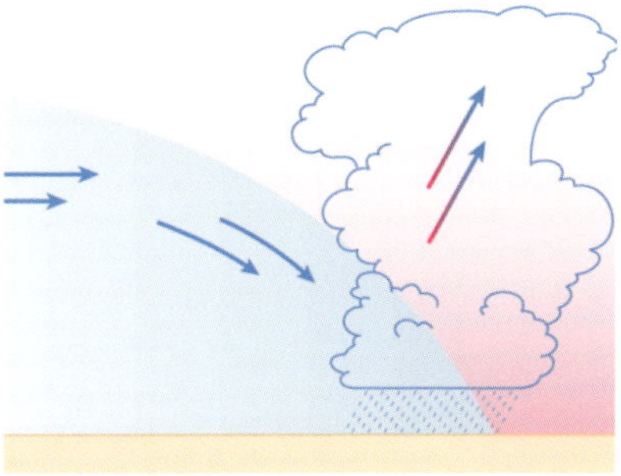

Front precipitation – cold front

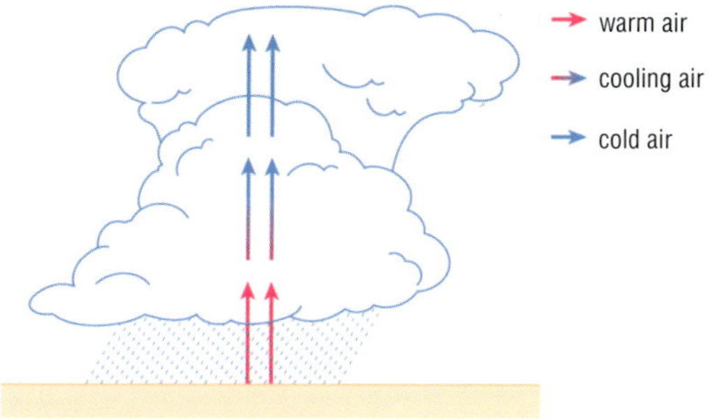

Convection precipitation

Barrier cloud precipitation

The principle of barrier cloud precipitation is quite simple: relatively warm, moist air masses stream against an obstacle, e.g. mountain flanks. Since the air cannot flow through the mountains, it has to climb. As it does, it cools off more and more. However, since cold air can absorb less moisture than warm air, at some point condensation occurs (water drops form), clouds take shape and, as a consequence, precipitation.

Front precipitation

Front precipitation is the consequence of warm or cold fronts. In a warm front, relatively warm air masses are thrust on top of the cold air masses, are thereby uplifted, condense and this leads to precipitation which is usually persistent. In a cold front, the heavier cold air masses slide like a wedge beneath the warmer air masses and elevate them abruptly. This precipitation usually strikes with intensity, but ordinarily does not last for long.

Convection precipitation

Convection precipitation is of relatively little importance to avalanche science. This occurs only when, in warm weather, the water condenses in the ground (without snow), rises, then, when it reaches saturation, forms clouds. releases precipitation. A typical example: thunderstorms in summertime.

Following heavy rainfall, the snow mass forfeits its firmness and fractures.

☐ Rain as an avalanche trigger

Rain-on-snow leads to an increase in avalanche danger for a short spell. On the one hand, the water seeps into the snowpack and creates an immense burden of weight. On the other, the rain destroys the bonding of snow crystals, acts like a lubricant. The latter effect takes place especially when rain seeps down to more deeply embedded layers inside the snowpack and is blocked by hardened layers from flowing further. That is when a water-saturated layer of several centimeters thickness can form. The direct consequence is usually a pronounced, naturally triggered avalanche.

☐ Melting metamorphism

The fundamental processes of snow metamorphism are described in dp.5 (snowfall after long period of cold).

A special form of this transformation is so-called melting metamorphism. For this to occur, the temperature of a snow layer must reach 0°C. The energy required for this can come from rising air temperatures; from solar radiation; or from rainfall. When the temperature of a layer of snow reaches 0°C, it cannot rise any further. By definition, snow can never reach temperatures above zero. That means any additional energy input goes towards the melting of the snow. A snow-water mixture is

Melting metamorphism under a microscope.

Water runoff finds its own path, usually in the form of canals which then freeze at colder temperatures.

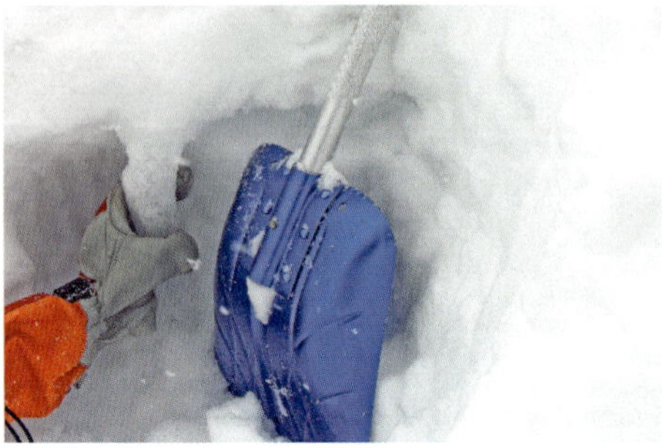

always precisely 0°C, and stays at that temperature for as long as it takes for the very last snow crystal to melt. Through the melting process, the snow crystals round out and the melt water which is freed fills the pores between the rounded snow crystals. This process is linked to the settling of the snowpack. To start with,snowpack firmness can even increase slightly; however, as the melting continues, i.e. when rainfall is strong or persistent, it quickly is lost.

☐ The positive influence of rain

Medium to long term, rain can also have a stabilizing effect, namely when temperatures drop following the rainfall; also during the cooling which occurs at night. In such cases, the free flowing water freezes in the pores and the snowpack consolidates. Depending on exposure to the cold, very thick melt-freeze crusts can form with the typical frozen-together, glassy, round grains of snow. Frequently, frozen melting canals inside the snowpack can be observed. This shows how the water is finding its way towards the ground. ▌

dp.3 practical tip

In case of heavy rainfall, stay home or at the refuge.

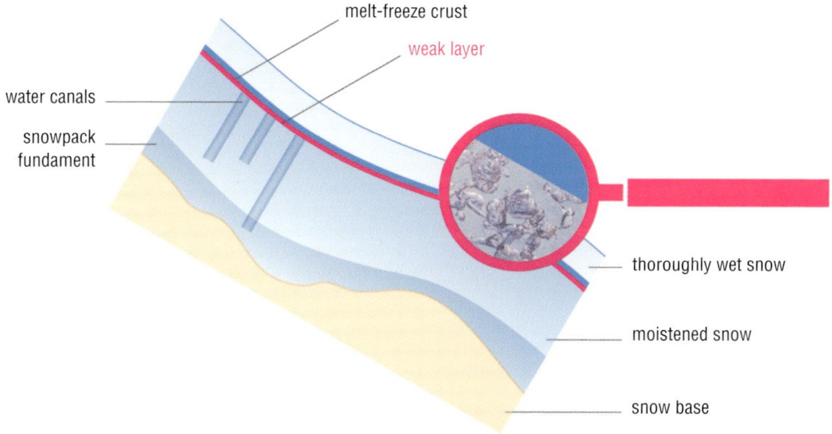

avalanche thurn.

Avalanche. During a period of unusually intensive precipitation on 31 January, a homeowner in Thurn near Lienz noticed in the stream bed adjacent to his house how snow masses which slid into the brook were accumulating. He made an attempt to remove this congestion and was caught by a mass of snow-and-water flowing upstream and swept along in the narrow stream bed for about 100 meters. He caught and stopped himself at an obstacle which traversed the stream, and was covered by 1.5 meters of water-saturated snow. The rescue squads went into action, but lost the race against time.

Short analysis. The avalanche is not directly related to rain, but to the effects of massive water seepage in the snowpack. "Slush" is the English word for this type of avalanche, which is extremely rare in these parts. It is a water-saturated snow mass which can be triggered into movement even by a tiny impluse on slopes with very low angular gradients. Subsequently, much like a mudslide, a destructive mass of water-and-snow develops. This phenomenon can often be observed in the wetlands of the high plateaus in Scandinavia where each spring, water-saturated snow masses form and accumulate, and move towards the valley floor. In this specific case, the water/snow mixture formed from several snow dams on the bed of the stream, caused on the one hand by heavy snowfall, and on the other, by snow masses which were dumped into the stream by snow removal squads. After one of these snow dams broke, a series of floodwaves developed. One of these took the homeowner by surprise.

spot of burial

Where Thurn / central East Tirol / 850 m / stream bed / 10°
Who 1 involved person / 1 fatality
When 31. 1. 2014, 9:15 am
Avalanche wet snow – slush / W 1 m / burial in 1.5 m
Regional danger level 4 (high)
Aval. bulletin headline High avalanche danger in East Tirol, southern Stubai Alps, soon in the Ötztal and Zillertal Alps
Avalanche problem wet snow

danger pattern.4
cold on warm / warm on cold

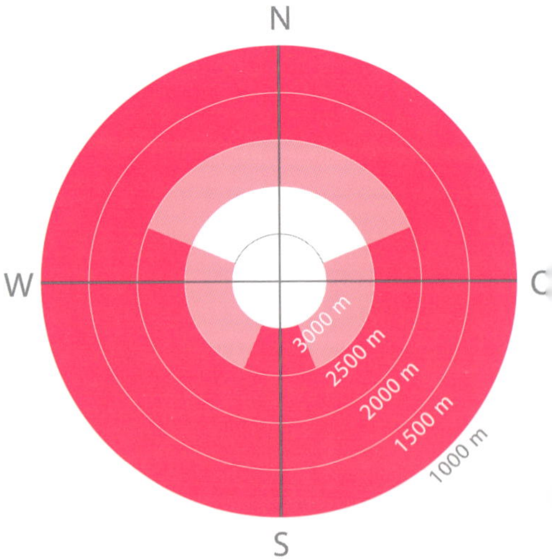

A theory purporting that huge temperature disparities during snowfall (regardless whether cold on warm or warm on cold) exercise a positive effect on the avalanche situation held sway for much too long in avalanche science. In fact, this is the case only under certain very specific prior conditions. For the most part, such temperature differences have a negative effect, since they reinforce snow metamorphism inside the snowpack. As a rule, a thin, weak layer forms which is highly prone to triggering. This layer is often found on south-facing slopes. It is highly treacherous for the additional reason that the weak layer does not yet exist immediately after the new snow is deposited, it forms only over the course of the following days.

avalanche urgtal.

The north-facing slopes of the Urgtal are a popular destination for freeriders of the Serfaus-Fiss-Ladis ski area. Hardly surprising, then, that on 7 April numerous skiers took advantage of the splendid weather and fresh powder snow to enjoy wide-sweeping turns. Through the impulse of one of the descending skiers, an initially small-sized slab avalanche was triggered which subsequently developed into an immense avalanche.

□ Circumstances of the accident

Just before 3:00 pm, a guided four-man group skis into freeriding terrain towards the Urgtal near the mountain terminal of the Almbahn lift. The route leads over sections of rolling knolls. At a flattening of the slope, the group stops to permit each group member to descend the slope singly. When one member of the group skis off, a small slab avalanche is triggered. The person is caught in it, swept along and comes to a stop after about 40 m when the ABS backpack inflates, uninjured and only partially buried. However, through the additional burden of the slab, a deep persistent weak layer is dislodged which then is released in a fracture of enormous size. An avalanche develops which is 850 meters wide and 1400 meters long(!).
During the avalanche there are about 20 people in the danger zone. Due to the complex situation and unusual size of the avalanche, a huge rescue effort is launched, with four helicopters, 150 searchers and avalanche dogs. Miraculuiously, no other persons were caught in the avalanche.

The avalanche is triggered at Point 1 and continues to grow from there through the successive numbers.

Where Urgtal / Silvretta-Samnaun / 2520 m / N-slope / 35°

Who 4 directly involved, several indirectly involved persons

When 7. 4. 2015, 3:00 Uhr pm

Avalanche slab avalanche (dry) / L 1400 m / W 850 m / fracture 0.2–1.5 m

Regional danger level 3 (considerable)

Aval. bulletin headline Frequent naturally triggered avalanches on sunny slopes

Avalanche problem new snow / persistent weak layers

The avalanche ultimately fractures down to a highly pronounced, deep persistent weak layer. This explains the magnitude of the avalanche.

Analysis

Weather and snowpack. What is most important in this accident is to review the weather in the two weeks prior. It was marked by ongoing and quite mercurial shifts of cold and warm air masses starting on 26 March. The 26th is significant because intensive, diffuse radiation and high air moisture gave the decisive impulse for a continuous moistening of the snowpack, even in high altitude shady terrain. Immediately thereafter, snowfall at very low temperatures set in. Variable conditions followed over the Easter holidays as warm and cold fronts plus bouts of heavy snowfall moved through, accompanied by storm-strength winds. The days immediately prior to the avalanche were like deep winter, seen from the point of view of temperature. Up to 30 cm of snow fell at low temperatures with wind impact.

On 7 April, then, cloudless skies, unimpeded sunshine. Snow analysis has shown that starting on 26 March, a thin layer of faceted snow crystals began to develop beneath a thin melt-freeze crust. This was at an altitude between about 2500 and 2900 m, particularly on shady slopes. However, stability tests can rarely pinpoint potential fracture propagation points inside a weak layer. Furthermore, a deep persistent weak layer from early winter was also found, rather typical for winter 2015, which in the interim – quite similar to the weak layer from 26 March referred to above – was not at all sensitive to triggering. On the snowpack surface, however, there were freshly deposited snowdrifts in

This photograph was taken on 28 March 2015. Two days prior, it was warm, the snowpack became moist. Then temperatures dropped and snowfall set in. The melt-freeze crust which is visible in the zone of the photograph developed a weak layer which played a role in the accident in the Urgtal.

dp.4 recognizing it

Without precise observation of the weather it is not easy to recognize this pattern. One needs to focus particularly on the shift from a warm period to a cold one, or vice versa. In each case it is a matter of whether snow layers of highly disparate temperatures are placed against one another. If that is the case, snow metamorphism can set in and a critical weak layer of faceted snow crystals can form (see dp.5). Whether the weak layer then grows to sufficient size to become a danger to skiers and freeriders must be gleaned from the avalanche bulletin or verified from snow analysis.

places which were buried by fresher powder snow. It seems most likely that the snowpack broke primarily inside the thin weak layer which developed as of 26 March. Through this "initial ignition" and the very heavy additional loading of the first avalanche, the area-wide deep persistent weak layer was massively disturbed.

Avalanche. During the Easter holidays which preceded the accident, several avalanches triggered naturally, some of them large-sized, due to the heavy snowfall, although not in the bowl of the Urgtal. Thereafter, the situation quieted down rapidly. For that reason, only the dangers of small-sized freshly formed snowdrift accumulations and loose-snow avalanches on sunny slopes were pointed out in the avalanche bulletin of 7 April.

Terrain. The initial avalanche is triggered at the orographic right edge of the major avalanche. This is very steep, north-facing terrain. With a high degree of probability, the skier at the time of triggering was in a spot where the snow was shallow, and where his sole impulse had a larger effect on the existing weak layers. Danger pattern 4 often occurs in very strictly limited altitude bands and aspects. Presumably the conditions for a weak layer forming in the fracture zone were nearly perfect. ▌

background knowledge: cold on warm / warm on cold.

dp.4 definition

The terms "cold on warm" and "warm on cold" mean that due to stark temperature changes between two phases of weather, the temperature of the already deposited snowpack and that of the newly arriving fresh snow are significantly different (more than 5°C disparity).

What happens inside the snowpack?

What is decisive is the same in each case: a huge temperature disparity inside of just a few centimeters of the snowpack stimulates snow metamorphism, the transformation of snow crystals. Potential weak layers are given the starting impulse to form.

Melt-freeze crust

Ordinarily the combination of cold on warm or warm on cold is connected to the formation of a melt-freeze crust. Either a relatively warm, moist snowpack surface gets encrusted by the swiftly dropping temperatures or a cold snowpack surface is moistened by rising temperatures or wet snowfall/rainfall, thereby leading to the formation of a generally thin melt-freeze crust.

Formation of weak layers

There is a widespread opinion that melt-freeze crusts are dangerous because new snow deposited on them bonds poorly with the crust, thereby increasing the danger of avalanches. Long experience and the analysis of hundreds of avalanches, however, show that a different process is responsible for heightened avalanche danger following a pronounced temperature change. Namely, in the

What is decisive is the shift from warm to cold or cold to warm: after a warm front has passed through, even the ice begins to melt.

borderline zone of a melt-freeze crust, a thin, weak layer usually forms composed of faceted snow crystals. If the snowfall is cold and the new snow is deposited on a relatively warm old snowpack surface, the weak layer ususally forms beneath the melt-freeze crust. Whereas if the snowfall is relatively warm and new snow is deposited on a cold old snowpack surface, formation of faceted snow crystals above the melt-freeze crust is the result. What is responsible for this is the transport of water vapor resulting from vapor pressure gradients.

Water vapor pressure is higher in relatively moist, warm snow than in dry, cold snow. Thus, vapor transport of warm, moist air to cold, dry air takes place. If then bonded snow (wind-drifted snow, settled new snow) is deposited on top of this melt-freeze crust, either the fresh snow glides on a thin, weak layer onto the melt-freeze crust or the fresh snow including the melt-freeze crust dissolves in a layer consisting of loose snow crystals which lies beneath.

□ A delicate matter

This is delicate, indeed, treacherous. Because immediately after the snowfall blankets the melt-freeze crust, no weak layer yet exists. It forms only during the course of the next few (typically 2) days. That means the avalanche situation actually worsens invisibly and unrecognizably without being directly caused by any subsequent weather conditions or other external effects. ▌

dp.4 recognizing it

This danger pattern is also treacherous because it occurs especially in sunny terrain, where ordinarily the snowpack layering is more stable than on shady slopes, and where statistically far fewer avalanche accidents occur. For that reason, skiers and freeriders give far less consideration to avalanche dangers there.

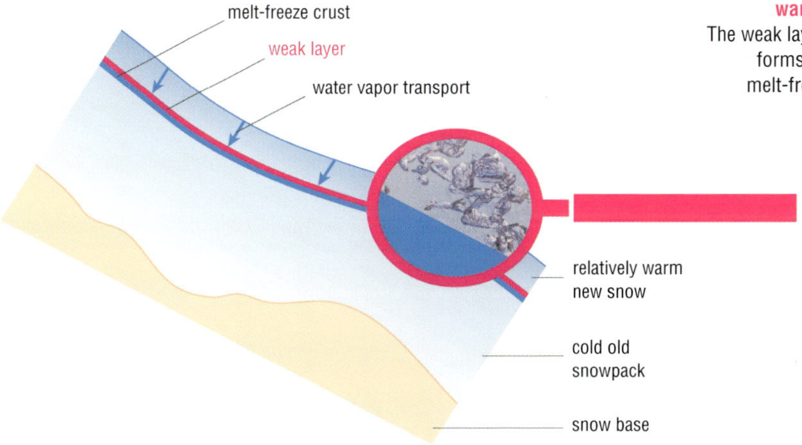

warm on cold
The weak layer usually forms above the melt-freeze crust.

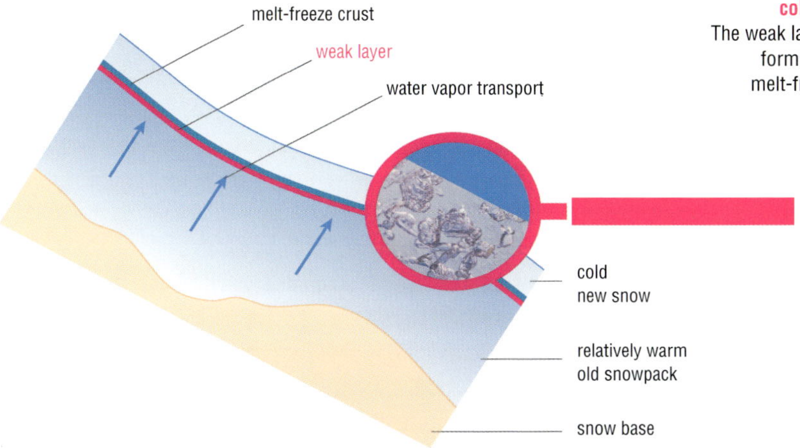

cold on warm
The weak layer usually forms below the melt-freeze crust.

avalanche hornköpfl.

Avalanche. Fresh snowfall entices skiers and freeriders to an attractive area near the Kitzbüheler Horn. Two locals traverse a very steep northeast-facing slope beneath the Hornköpfl refuge. One of them waits for the other on the slope. When the second skier joins him, a slab avalanche is triggered and sweeps both of them along, 170 m and 400 m respectively. The second person is only partially buried and is able to free himself from the snow masses with only slight injuries, by the time the rescue squad arrives. The other person is pinpointed with an avalanche transceiver and freed from 2 meters of snow masses. He is resuscitated and flown to the hospital, where he dies on 30 January.

Short analysis. A thin, treacherous weak layer which formed a few days prior to the accident is the obvious, direct cause of the avalanche. It was foggy and warm at the time. The snowpack surface was moist. Thereafter, temperatures dropped and a thin melt-freeze crust formed, beneath which another thin layer consisting of faceted snow crystals developed. When there was snowfall on the night of 27-28 January (30-50 cm, accompanied by heavy winds) the weak layer is “armed” by that event. That means the snowdrifts deposited on top of it created new tensions which could be transferred over additional surface areas. These snowdrifts caused the lift company in Kitzbühel to unleash explosions in the early morning hours in order to secure the ski area. As a result, fresh slab avalanches could be observed in the vicinity of the ski area on the day of the accident. These artificial triggerings should have served as indicators of potential avalanche danger, even if the “culprit” weak layer itself could only have been discovered by professional avalanche scientists in the days which followed.

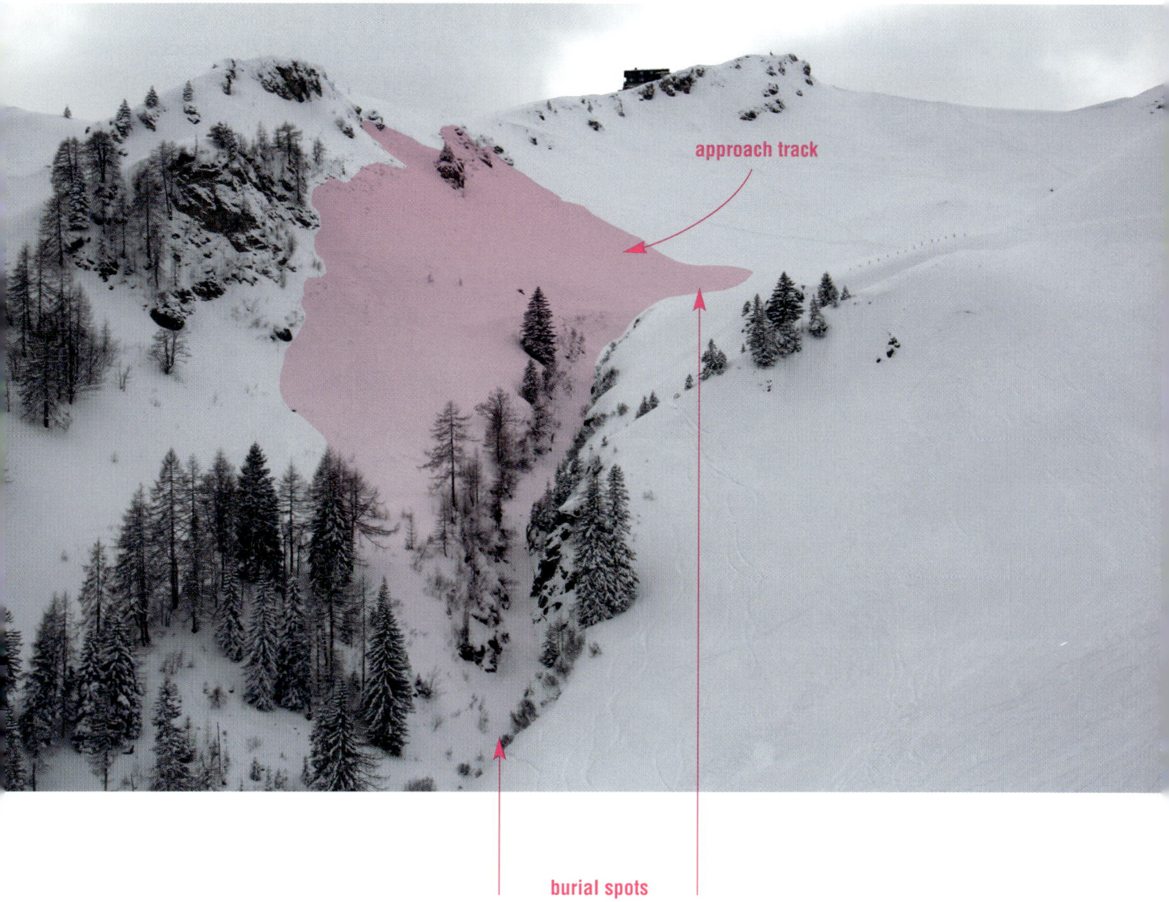

Where Hornköpfl / Kitzbühel Alps / 1690 m / NE-slope / 40°

Who 2 persons involved / 1 fatality, 1 injured person

When 28. 1. 2015, 2:10 pm

Avalanche slab avalanche (dry) / L 400 m / W 150 m / fracture 0.65 m / burial 2 m / appx. 30 min.

Regional danger level 3 (considerable)

Aval. bulletin headline New snow and wind creating a treacherous avalanche situation

Avalanche problem new snow

avalanche hochschober.

Avalanche. Three experienced backcountry skiers climb through the Ralftal to the Hochschober. The ascent is unproblematic. They deposit their skis and one person remains behind. When the others return from the summit, they discover an avalanche fracture between the spot where the skis were left and their ascent route. The other person initially waited for their return, then moved onward in the direction of the summit and was caught by the avalanche and totally buried. He didn't survive.

Short analysis. The accident analysis carried out with the Alpine Police revealed an unbelievable chain of improbable circumstances. It turned out that one of the skiers who continued on to the summit selected a slightly different descent route when he returned to the spot where they left their skis. Soon, however, he noticed that the terrain was extremely steep and decided to retrace his steps and return to the ascent route. Doing this, he triggered a tiny slab avalanche just beneath his track and did not notice it. Subsequently, another slab avalanche releases just below the first one. The person waiting at the spot where the skis were left was in the immediate vicinity, at the edge of the second avalanche which triggered, and is caught in its flow and buried by it at the bottom of the slope. Stability tests showed a varyingly pronounced thin layer of faceted snow crystals near the melt-freeze crusts. The repeated moistening of the snowpack by solar radiation in extremely steep, wind-exposed, east-facing terrain, together with the subsequent period of bad weather, explains this snowpack layering. The likelihood of triggering, however, was not very great. What was decisive was the extreme steepness of the slope. ▌

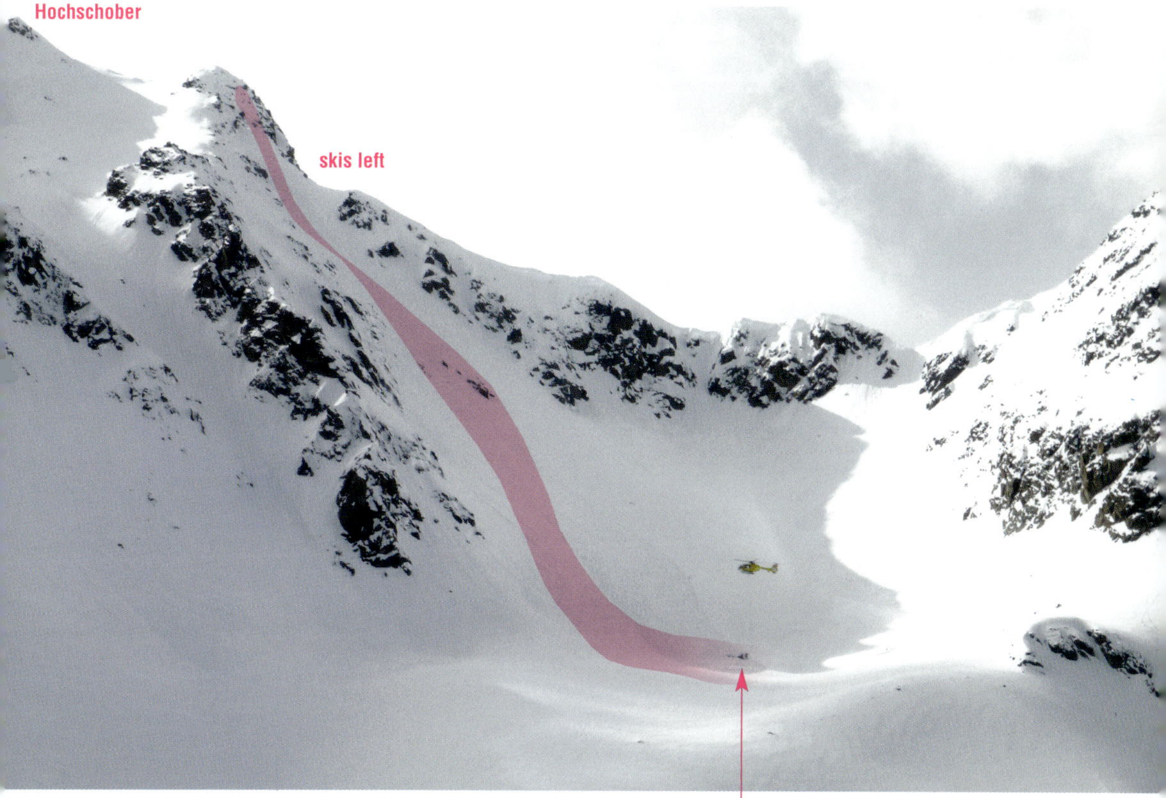

Where Hochschober / central East Tirol / 2850 m / E / 45°

Who 3 persons involved / 1 fatality

When 13. 4. 2013, 3:20 pm

Avalanche slab avalanche (dry) / L 250 m / W 10 m / fracture 0.3 m / burial 0.5 m

Regional danger level 2 (moderate)

Aval. bulletin headline Significant increase in avalanche danger – Beware wet loose-snow avalanches

Avalanche problem new snow / persistent weak layers

danger pattern.5
snowfall after long period of cold

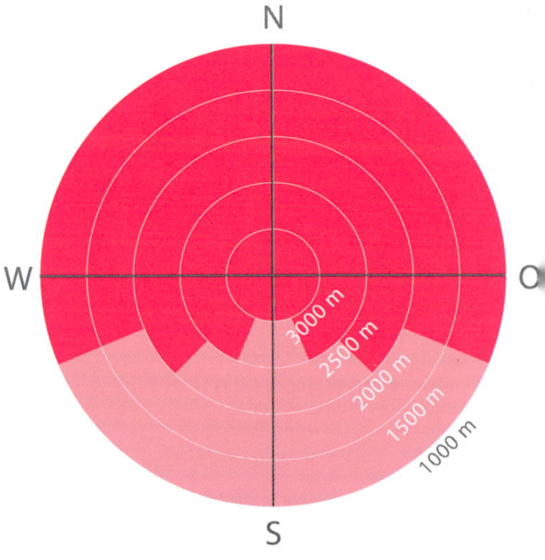

Dec Jan Feb

Among avalanche events, this is a classic: following a lengthy period of cold, it begins to snow. In addition, strong winds prevail which transport the new snow accordingly. Within the shortest imaginable time, a treacherous situation arises for backcountry skiers. This is also the case if, following a lengthy period of cold, "only" strong winds are blowing without snowfall. The problem is that fresh snowdrifts are deposited on wind-protected leeward slopes on top of a loose old snow cover surface which usually contains depth hoar. Wind-drifted snow and persistent weak layers are very poorly bonded with each other. The snowpack is just waiting to be triggered by additional loading.

avalanche hochgurgl.

Admittedly, it is not always easy to resist the call of skiing through untracked terrain after fresh snowfall and beautiful weather conditions which frequently follow. Nevertheless: whenever it is a matter of dp.5, iron discipline is the most important survival strategy. One local freerider was enticed by the beautiful new snow to ignore obvious signs of danger. It unfortunately cost him his life.

□ Circumstances of the accident

Around midday in Hochgurgl, two avalanches are reported. The helicopter stationed in Hochgurgl takes off on an exploratory flight. The crew notes in one of the conical-shaped avalanche paths quite near the helicopter base two skiers lying atop the avalanche. A rescue member is immediately lowered and begins his search with an avalanche transceiver. He pinpoints a signal and can quickly free a buried skier. It is a local, well-equipped freerider who left red ski run no. 20 around midday and descended the very steep slope.

His spot of burial is located directly on the closed Timmelsjoch High Alpine Road.. The abrupt change in slope gradient in the avalanche path gave rise to deeper burial than would have occurred, for example, in a more regularly dropping avalanche path. The person is resuscitated and flown to Innsbruck Hospital, where he dies.

Where Hochgurgl / southern Ötztal Alps / 2500 m / NW-slope / 35°

Who 1 involved person / 1 fatality

When 16. 2. 2012, ca. 11:00 am

Lawine slab avalanche (dry) / L 260 m / W 20 m / fracture up to 0.6 m / burial appx. 1.5 m / appx. 60 min.

Regional danger level 4 (high)

Aval. bulletin headline High avalanche danger widespread above 1600 m!

Avalanche problem persistent weak layers / wind-drifted snow

Analysis

Weather and snowpack. One day prior to the accident, a two-week period of cold came to an end, during which time the snowpack surface had starkly transformed. It consisted of unbonded, faceted snow crystals; of depth hoar; and to some extent of surface hoar. This is a mixture of the deadliest weak layers a snowpack can have. The old snowpack beneath this layer was stable, so that a fracture down to more deeply embedded layers was unlikely. When on 15 February heavy snowfall set in amidst strong winds, avalanche danger suddenly surged. With about 40 cm of new snow in the area of the accident, extremely trigger-sensitive snowdrift accumulations formed in all aspects. The danger for a trained eye, despite the improved weather conditions on 16 February, was obvious. Apart from naturally triggered avalanches, increasingly frequent glide cracks appeared and settling sounds were heard ongoingly in outlying terrain.

Avalanches. Within just a few days, 50 avalanches are reported, 17 on the accident day alone. The significance is clear. Once again, dp.5 is confirmed as the most accident-prone danger pattern of all. The critical situation does not come as a surprise and is reported in all the media. In the blog of the Avalanche Warning Service on 14 February one can read: "The combination of unfavorable snowpack layering, snowfall and storm-strength winds will give rise to a significant increase in avalanche danger as of 15 February! Avalanche danger will presumably increase further to danger level 4 (high)

dp.5 recognizing it

At least once every winter, a lengthy period of cold sets in, usually a number of such periods. This situation does not escape the notice of skiers and freeriders. There is often sunshine during the cold period. Loose powder snow or faceted snow crystals spray wide swaths on their descents; avalanche danger towards the end of such a period is usually low. What distinguishes this danger pattern is the abrupt increase in danger as soon as snowfall sets in amidst wind impact, or as soon as strong winds arise. These are weather phenomena which should not escape the attention of any skier or freerider. The problem is that during a period of cold, a slope which can skied without worry suddenly (in a few hours) turns into a deadly trap.

Naturally triggered avalanches are visible in this slope sector in Hochgurgl. Including natural triggerings between avalanche barriers.

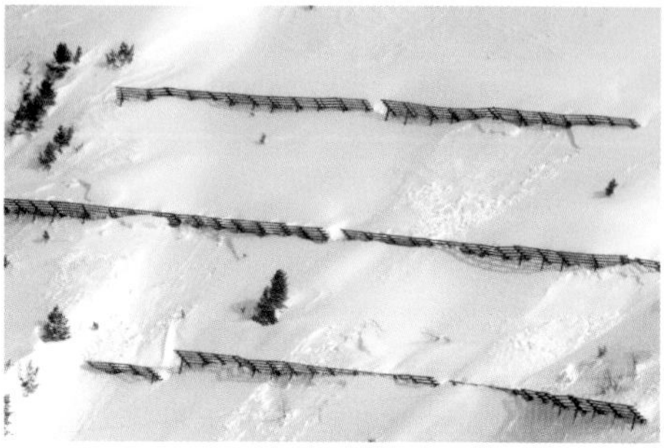

during the afternoon of 15 February. Slab avalanches can then be triggered quite easily by skiers and freeriders. We also expect naturally triggered avalanches, not merely in terrain adjacent to ridgelines, but also below the treeline...."

On the day of the accident, high avalanche danger prevails above 1600 m over widespread areas. Conditions in Tirol's backcountry touring regions are judged to be critical. The Avalanche Warning Service calls urgent attention to the fresh snowdrifts in all aspects. Since knowledge of avalanche signs and dangers is imperative to all skiers and freeriders in outlying terrain, the inexperienced are urgently advised not to leave the secured ski runs. You can see numerous fresh, naturally triggered, but also artificially triggered avalanches immediately adjacent to the ski runs, even dotting the zones between ski barriers.

Terrain. The accident slope has a gradient of 35°, faces northwest, is full of bowls and mogels, permitting easy deposition of snowdrifts. What stands out is the heavy wind impact: in spite of 40 cm of new fallen snow, wind-exposed zones are bare of snow, completely windblown. Next to them are concavities utterly full of snowdrifts. These are just as visible as the naturally triggered avalanches. In short: the descent route has not been selected well, i.e. in accordance with the dangers present. ▌

On the day of the accident, high avalanche danger prevails in Tirol. The upshot: numerous naturally triggered avalanches.

background knowledge: snowfall after long period of cold.

dp.5 definition

In meteorology there is a distinction between icy days and frosty days. An icy day occurs when temperatures remain below freezing all day long. A frosty day occurs when the temperature drops at most to the freezing point. Lengthy periods of cold, i.e. several icy or frosty days in sequence, give rise to metamorphism inside the snowpack, which then has a negative effect on the avalanche situation as soon as snowfall occurs.

The metamorphism

Snow is a highly complex material which is constantly undergoing change. Snowflakes are the term used for several interconnected snow crystals. Even during snowfall, spokes and branches break away from the symmetric, hexagonal snow crystals. Wind and temperature shifts within the various layers of the atmosphere through which the snow crystals descend effect changes as well. When the snow crystals ultimately reach the ground, the first actual process of transformation begins. It is known as snow metamorphism.
It is necessary to distinguish between equilibrium and kinetic metamorphism. A different, special form is melting metamorphism, described in dp.3 in detail. Through these processes of transformation, different forms of snow crystals and different sized snow crystals form. In addition, the density, proportion of pores and the structure – all of which are significant components of the snowpack – also undergo change.

Equilibrium (destructive) Metamorphism

A completely normal and regular hexagonal snow crystal (= new snow) falls to the ground and as a result of mechanical forces (crystals crashing to the ground, colliding with each other, wind influence) several of its arms and branches break off. The snow crystal becomes smaller. This crystal

Equilibrium (destructive) metamorphism: from a hexagonal crystal of new snow to decomposed forms to the final product of the transformation, a rounded grain.

is called decomposed snow. This decomposed snow still has numerous lateral arms. But since under the laws of nature such branches are unfavorable as regards energy consumption, a more favorable form is sought for purposes of energy efficiency: a rounded, spherical grain. In other words, attaining maxi-mum volume with minimum outer surface.

That means ongoing transport of mass takes place from the branches of the snow crystal towards the center. The crystal turns more and more into a sphere. In the process, great loss is incurred again (thus, it is destructive metamorphism), the space occupied by pores inside the snowpack is reduced, the snowpack settles and consolidates because the grains come into closer contact with each other. This process is heavily dependent on the temperature of the snow: the closer it is to 0°C, the faster it proceeds. At lower temperatures, the destructive process slows down. As a rule, the destructive transformation takes several days. But in unfavorable conditions, e.g. very low temperatures lacking solar radiation, it can take up to two weeks and longer.

□ Kinetic (faceting) metamorphism

Deep persistent weak layers seldom have temperatures below about –1°C to –2°C (in glacier ice the temperatures can be well below that). Due to these relatively mild temperatures, compared to the snow layers closer to the upper surface, the water vapor rises from ground level towards the upper,

Kinetic (faceting) metamorphism: from rounded grain to faceted crystal to cup-shaped crystal. At right, the faceted layer buried by snowdrift is visible.

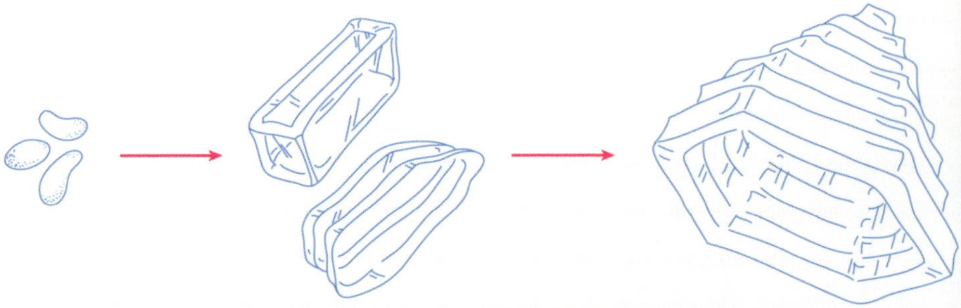

colder layers of snow and here sublimates, i.e. again becomes solid. Thereby, snow crystals again form (this time, kinetic metamorphism) and take on angular, polygonal shapes. Since these polygonal crystals are no longer well bonded to each other (fewer contact points than with spherical grains) firmness and cohesiveness of the snowpack decrease.

The final forms in kinetic (faceting) metamorphism are so-called cup-shaped crystals. They are large concave crystals (often several mm large). A layer of cup-shaped crystals is known as depth hoar. This process ordinarily takes somewhat longer than equilibrium (destructive) metamorphism. Two to four weeks until cup-shaped crystals are ultimately formed is typical. Since this process is unrelated to a settling of the snowpack (in contrast to equilibrium metamorphism) it is not recognizable.

The decisive criterion for kinetic metamorphism of snow crystals is the temperature gradient inside the snowpack. In the graph below, the major difference can be seen: if other conditions remain the same, i.e. ground temperature of -1°C and air temperature of -11°C, you get, depending on snow depth, completely different gradients. If snow is shallow (10 cm deep) you get a gradient of 10°C per 10 cm. If the snow is deeper (1 meter) the temperature gradient is just one tenth, namely, only 1°C per 10 cm. The less snow there is, and the colder the air is, the larger the temperature gradient inside the snowpack is. And the larger the temperature gradient, the faster the steps of kinetic metamorphism proceed.

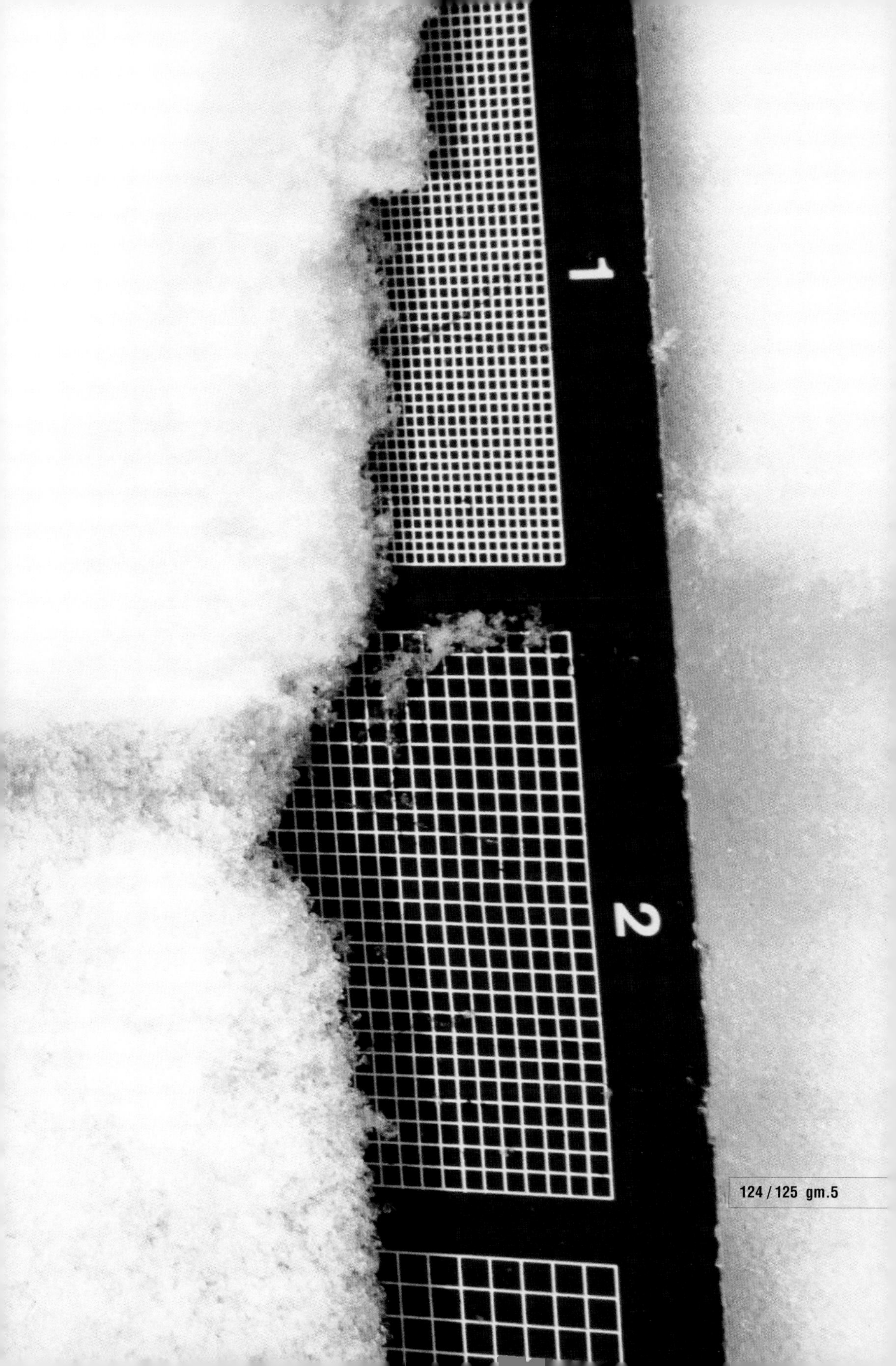

124 / 125 gm.5

A lengthy period of cold can give rise to the formation of weak layers at different places inside the snowpack.

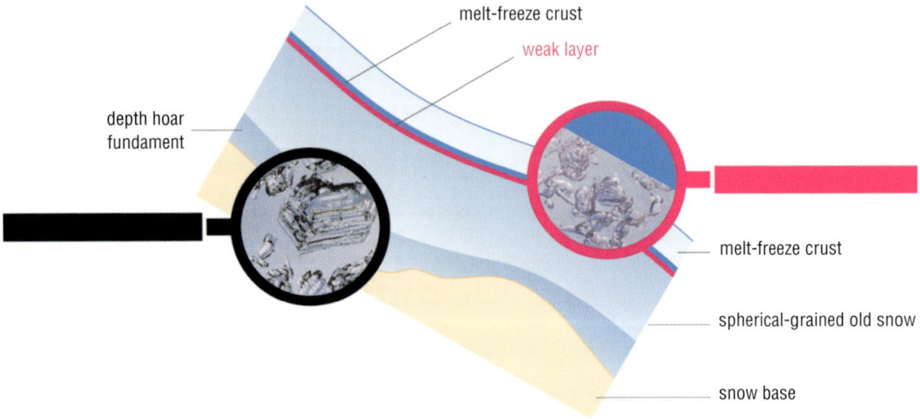

If the entire snowpack has been kinetically transformed, that does not mean it is dangerous. Quite the contrary: that is a snowpack nearly free of tensions, filled with the very finest grains of "powder snow."

It becomes critical only when this loose, unbonded fundament is buried by fresh snowfall accompanied by wind influence. Depth hoar, i. e. faceted snow crystals, can then also be found beneath hardened layers of wind crusts or melt-freeze crusts. That explains why particularly in cold winters with shallow snow, such as in 2009/10, the snowpack layering was so poor and the avalanche situation so critical. ▌

dp.5 practical tip

Adapt your behavior swiftly when the avalanche situation changes following a period of cold.

If external conditions are the same, the temperature disparity inside the snowpack is much higher when the snow is shallow than when the snow is deep.

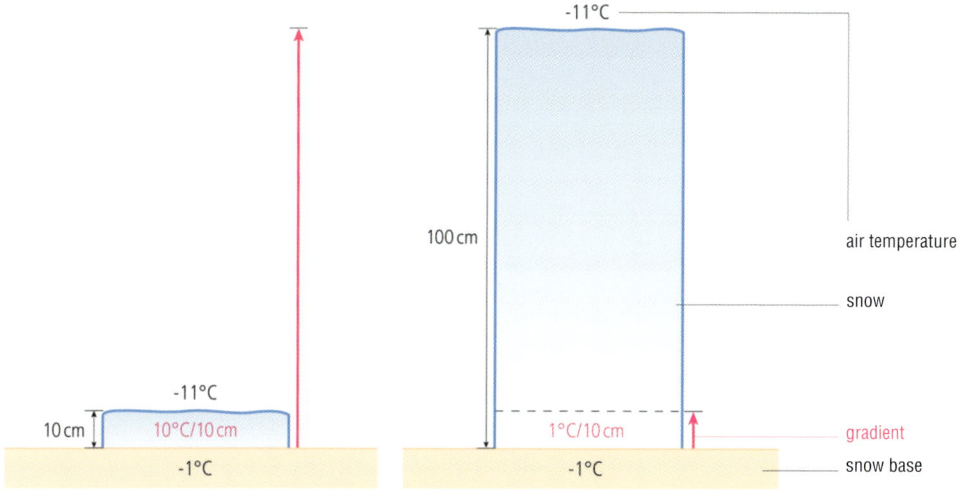

1 new snow 2 decomposed snow 3 round grains 4 faceted grains 5 cup-shaped grains 6 melting shapes

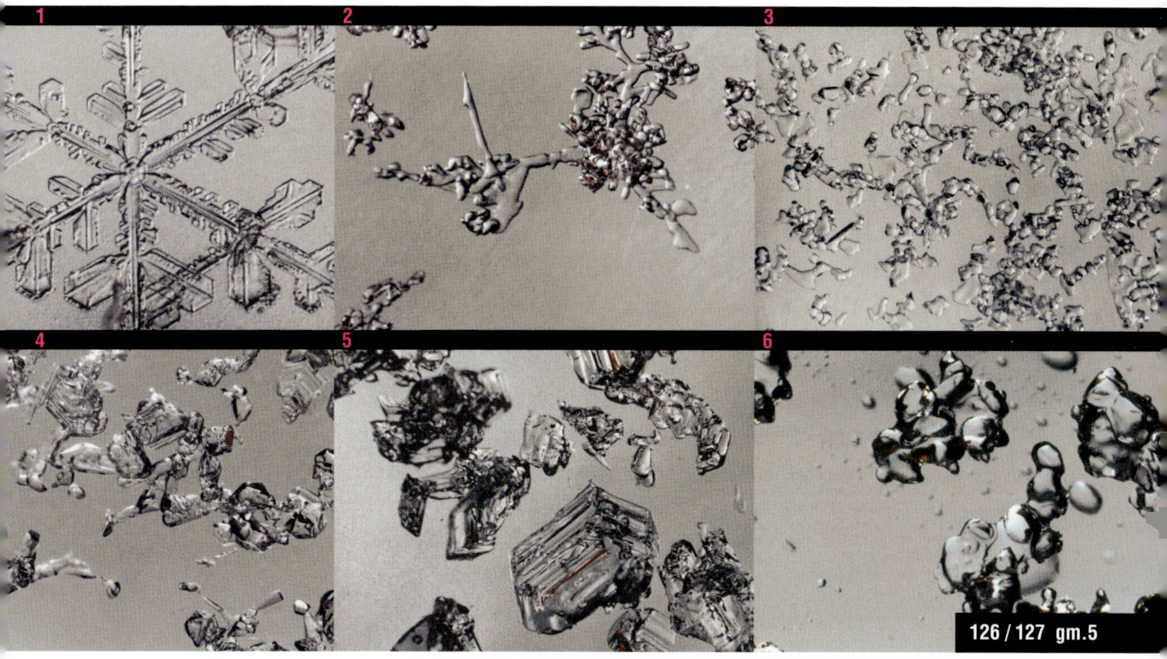

avalanche grosser gamsstein.

Avalanche. A married couple climbs from the parking lot of the Hochfügen ski area up to the Sonntagsköpfl without any problem. On the summit they decide to take a different descent than the ascent route. They traverse extremely steep, east-facing terrain below the Grosser Gamsstein, where they trigger a slab avalanche and are completely buried in the snow masses. The accident is not noticed by anyone. The man is still conscious when the avalanche comes to a halt, for he has a large oxygen hole around him. After about 5 hours, at 7:02 pm, he manages to send an emergency call from his spot of complete burial and to direct the helicopter to the spot of the accident, despite approaching darkness. At 8:30 pm, hypothermic but unharmed, he is freed from the avalanche. In the interim, his wife has died.

Short analysis. It is an extremely delicate time for skiers and freeriders. The primary cause is found in the increasing cold which started on 20 January. In average snow depths, the existing hardened layers are incrementally reduced and the snowpack becomes looser and looser with each successive day. Starting on 28 January, snowfall sets in. Strong winds are blowing which lead to snow transport and fresh snowdrifts accumulating. Subsequently, a large number of avalanche accidents with fatalities occur throughout the Alps. A period begins in which avalanche danger is assessed as considerable, or even high. An additional indicator for unfavorable conditions: the rescue squad triggers yet another avalanche during its rescue operation. Fortunately, no casualties are incurred.

Where Gamsstein / Tux Alps / 1950 m / E-slope / 40°

Who 2 persons involved / 1 fatality / 1 injured person

When 4. 2. 2010, appx. 2:00 pm

Avalanche slab avalanche (dry) / L 150 m / W 40 m / fracture 0.3 m / burial 0.5 m / 5 hrs

Regional danger level 3 (considerable)

Aval. bulletin headline Naturally triggered avalanches possible this afternoon on extremely steep, sunny slopes.

Avalanche problem persistent weak layers

avalanche mölser berg.

Avalanche. A group of 12 persons bases their backcountry touring weekend at the Lizumer Refuge. On 5 February they plan an easy-going tour to Mölser Berg in superb weather. They decide to ascend the last section to the summit over the initially rolling, then increasingly steep east-facing slope. As seven of the participants arrive in the safe zone of the ridge and three others are just below them in a flattish spot, a very audible settling sound (WHUMPF!) is heard. A large slab avalanche then triggers which sweeps along two of the group about 150 meters. All the others react swiftly, activate their backpack airbags and are buried only partially. They remain unharmed.

Short analysis. It has been snowing in the Tux Alps since 31 January, about 60 cm of new snow has been registered, temperatures are low, winds are strong. The old snowpack on W/N/E facing slopes contains depth hoar over widespread areas, on shady slopes there is surface hoar. The formation of these layers was generated by the shallow snowpack in January, on the one hand; and by the arctic temperatures starting on 22 January, on the other. The freshly formed snowdrifts bonded poorly with the old snowpack. Several naturally triggered avalanches have been observed. The loose snowpack surface on day of the accident, formed by slackening winds at the end of the recent bout of snowfall, is particularly treacherous. Signs of wind are frequently buried by new snow. Espe-cially characteristic of this release is the wide-area spread of the avalanche due to the fundament of depth hoar. ▌

Where Mölser Berg / Tux Alps / 2440 m / East-facing slope/ 35°

Who 12 persons involved / 2 caught in the avalanche

When 5. 2. 2005, 11:30 am

Avalanche slab avalanche (dry-snow) / L 300 m / W 300 m / fracture depth 0.5 - 2.5 m

Regional danger level 3 (considerable)

Aval. bulletin headline Treacherous situation for backcountry tours!

Avalanche problem persistent weak layers

danger pattern.6

loose snow and wind

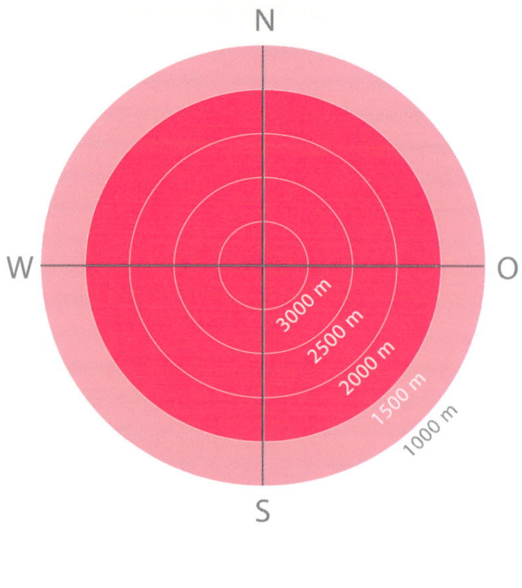

“Wind is the architect of avalanches.” This classic proverb from the 1930s by Wilhelm Paulcke remains unchanged even today. Wind influences both snow which is falling and snow which has already been deposited. It is a major factor in the formation of avalanches. When the snow is loose and dry, wind transports it, thereby increasing avalanche danger.

The colder the transported snow, the more sensitively it reacts to additional loading, because it is so brittle. It is characteristic of this danger pattern that the weak layer usually consists of new snow and is buried by snowdrifts. Thus, there must have been snowfall at low temperatures without wind impact either just before or in a previous period when the drifting began; or else the snowfall set in without wind influence, but winds arose during the snowfall. This is a pattern which ordinarily can be recognized with ease and lasts only for a short time. The exception is a situation in which the old snowpack surface consisting of loose, faceted crystals is itself transported by wind. When that happens, hard, brittle slabs form which remain prone to triggering for a lengthy period.

avalanche windachferner.

It is an everyday scene as has occurred thousands of times on the Stubai glacier: a skier takes a fall, the ski binding opens, the skier loses his ski. But this time, the fugitive ski unleashes a tragic chain reaction.

□ Circumstances of the accident

A 25-year old skier loses a ski when he falls on the Stubai glacier. The ski glides along the edge of the ski run out into free terrain and gets stuck in a bowl below a wind-carved crater. The skier tries to make his way to his ski, at first over a broad ridge, then across a small bowl. As he descends toward the ski, he triggers a small slab avalanche in extremely steep terrain adjacent to a ridgeline. He is equipped with an avalanche transceiver and is freed from the snow masses 30 minutes later. Attempts are made to resuscitate him, he is flown to Innsbruck Hospital, where he dies in the early hours of 7 March.

□ Analysis

Weather and snowpack. It is the beginning of March, ordinarily a time when the springtime situation begins and the typical daytime cycle of avalanche dangers is anticipated and taken into consideration.

spot of burial

Where Windachferner / southern Stubai Alps / 2950 m / SW-slope / 40°

Who 1 person involved / 1 fatality

When 6. 3. 2014, 3:00 am

Avalanche slab avalanche (dry) / L 30 m / W 30 m / fracture 0.2–1 m, burial 1.5 m / 30 min.

Regional danger level 2 (moderate)

Aval. bulletin headline Moderate avalanche danger widespread

Avalanche problem wind-drifted snow

On the day after the accident, new snowdrifts were well bonded with the previous loose snow.

The first week of March, however, is usually more heavily weighted towards the side of winter, particularly from the vantage point of temperatures. At high altitudes temperatures were -10°C and below. There was repeated snowfall during the days prior to the accident. A strong, cold northerly wind was blowing at high altitude.

The old snowpack is generally well structured. The problem is caused by the recently formed snowdrift accumulations. With a certain degree of experience, the danger is easily recognized in this terrain. On the one hand, you can see the freshly formed cornices at the edge of the bowl where the skier descended into it. The avalanche fractures at the borderline between recent fresh fallen snow and the denser snowdrifts which are deposited on top of it. Stability tests which were carried out during the following days show quite good bonding of these layers with one another. This confirms that such situations frequently last only for a very short time. The rising temperatures and solar radiation which prevail following the day of the accident also improve the situation.

Avalanches. The snowpack's proneness to triggering is not acute enough that slab avalanches can trigger naturally. Only the solar radiation, which intensifies during the course of the day, could give rise to isolated loose-snow avalanches in extremely steep, wind-protected terrain. The relevant danger zones are limited to very steep ridgeline terrain at high altitude. The avalanche danger level, moderate, corresponds to that situation.

Terrain. At the scene of the accident we find a classic trap where relatively shallow snow can nonetheless lead to deep burial in snow. Northerly winds have given rise increasingly to avalanche prone locations on south-facing slopes, corresponding to the scene of the accident. What is decisive for the triggering is the 40° gradient at the uppermost edge of the bowl. As is well known, the likelihood of avalanches triggering is ordinarily higher as gradients increase. But if you examine the terrain precisely, you can easily find safer alternatives to reach the lost ski, one of which was right next to the ski track itself, where the terrain is notably flatter.

In summary, with a certain degree of experience, the danger situation would have been recognized: fresh snowdrifts adjacent to a ridgeline in still quite low temperatures in terrain which is extremely steep and near a ridgeline. But let's face it, many skiers would have been more anxious to retrieve their lost ski and would have concentrated on that, instead of the dangers lurking nearby.

Recognizing loose and bonded snow. The easiest way to find out whether snow is loose or bonded is with a shovel. Simply fill the shovel with the snow layer and shake it slightly. If it falls apart right away, it is loose snow. If it doesn't, the snow is bonded, either through the process of settling or because it is slightly moistened. Loose snow cannot transfer tensions, bonded snow can. Even slightly bonded snow creates slabs ("soft slabs").

dp.6 recognizing it

As a rule, this danger pattern can be easily recognized, since it is found in the uppermost, wind-impacted layer of snow. A prerequisite for this, however, is the correct interpretation of wind signs – drifts behind protruberances, wind ripples or cornices – which can be grasped without difficulty. Their interpretation is more difficult in those rare cases when only light wind or no wind at all is blowing, and not at the onset but towards the end of snowfall if the snow was heavily transported during the major snowfall phase. In that case, loose snow can blanket recently formed avalanche prone locations. Nevertheless, there are usually wind signs evident elsewhere in open terrain. Otherwise, weather station graphs can provide assistance. Or a simple stick-test, with which bonded snow masses can be detected beneath the loose snow on the surface.

Recognizing the signs of wind. Wind activity is usually easy to spot in open terrain: cornices, wind ripples, wind dunes, drifts against protruberances or concavities all speak loudly and clearly.

What needs to be focused on every time: very often, one only sees the signs of most recent wind activity (for example in the early morning, at the door of the hut). These wind signs can be deceptive, however. Perhaps a front just passed through. Perhaps the wind direction shifted. Perhaps one perceives light NW winds during snowfall, with only moderate snow transport, without realizing that during the night, strong SW winds brought about massive snow transport. Conversely, pronounced signs of wind, e.g. large cornices, can sometimes be attributed to weather conditions which lie further back in the past, thereby falsifying your view of the current situation.

For that reason, you should always take advantage of the Avalanche Warning Service reports in addition to your own observations and interpretations. More than 100 automated weather stations with all possible wind data are available 24/7, free on the Internet. ▌

1 protruberance drift 2 cornice 3 wind ripples 4 snow plumes 5 concavities 6 tail of a comet

1

2

3

4

5

6

background knowledge: loose snow and wind.

wind definition

Wind is air in motion in the atmosphere which has two attributes: direction and velocity. The main cause of wind is found in disparities in air pressure. Air flows out of zones with higher air pressure (high pressure zones) into zones with lower air pressure (low pressure zones) in an effort to balance the disparity. Since such pressure disparities are being created ongoingly due to globally differing solar radiation, wind is ceaseless, never rests, never stops.

Wind direction

Wind direction is indicated either in degrees on a 360° scale or in a direction of the compass. 0° and 360° correspond to north; 90° to east; 180° to south; and 270° to west. Don't forget that the topography of the landscape also has an influence on wind direction: mountains and valleys, broad ridges and crests divert the wind, so that weather reports frequently release different data of wind direction from what is perceived by the senses. **Important:** direction is **where** der wind is coming from: a northerly wind comes **from** the north, does not blow northwards!

Wind velocity

Wind velocity is measured either in meters-per-second or often in Beaufort (Bft), since the effects of wind can thereby be easily assessed. Reported wind speeds are average values measured in 10-min-ute intervals. Frequently, values of top windspeeds (gusts) are also reported. That is the maximum value (for 2 seconds) within a 10-minute interval. Ordinarily, the values of wind gusts are approximately 1.5 to 2.0 times the average windspeed. Intensifying wind velocity with ascending altitude also requires attention: a strong wind can easily attain storm strength at a ridgeline or summit altitude. The following table facilitates recalculation of various wind velocities.

Wind velocities in various units and effects on snow transport.

Beaufort (Bft)	m/s	km/h	Effect
0 windless	0,0 – < 0,3	0	no air movement
1 slight breeze	0,3 – < 1,6	1 – 5	barely noticeable, no snow transport
2 light wind	1,6 – < 3,4	6 – 11	perceptible wind, no snow transport
3 weak wind	3,4 – < 5,5	12 – 19	snow transport only of extremely dry, light, fluffy snow
4 moderate wind	5,5 – < 8,0	20 – 28	snow transport begins
5 brisk wind	8,0 – < 10,8	29 – 38	audible wind, snow transport observable
6 strong wind	10,8 – < 13,9	39 – 49	audible whistle of wind, increasing snow transport
7 stiff wind	13,9 – < 17,2	50 – 61	trees bend, wind resistance while walking, widespread snow transport
8 stormy wind	17,2 – < 20,8	62 – 74	large trees in movement, walking is noticeably impeded, snow transport is increasingly irregular
9 storm	20,8 – < 24,5	75 – 88	maintaining equilibrium difficult, turbulence, irregular snow transport in all aspects
10 fierce storm	24,5 – < 28,5	89 – 102	ski tours generally not possible
11 gale-strength storm	28,5 – < 32,7	103 – 117	ski tours generally not possible
12 hurricane	> 32,7	> 117	ski tours generally not possible

Snow transport

At center stage of how wind creates avalanches is snow transport by wind. What matters most in the transport of snow is the type of snow crystals. If they are loose and dry, transport begins already at windspeeds of 15 km/hr; even lighter winds are sufficient during snowfall. Distinctions are made between snow-sweeping, which remains below eye level and does not impede visibility, and driven snow which reaches higher than 2 meters. Most snow is transported near the ground, up to about a half-meter above the surface. **Important:** snow transport increases disproportionately as wind velocity increases. The rule of thumb: doubling wind velocity transports up to eight times more snow! However, experience has shown that when winds are strong to stormy, i.e. blowing at 40 to 75 km/hr, wind direction remains more constant than at higher windspeeds. For that reason, it is at this windspeed that the most snow is transported into avalanche starting zones. At higher windspeeds, turbulence intensifies, which means the wind direction continually changes and thereby, fewer snow masses are transported, comparatively speaking. The amounts of snow transported by wind can be enormous. It is not unusual for a windward, upwind slope (facing the wind) to be utterly bare of snow, so that even grass and stones are visible; whereas a few meters over the edge on the leeward, wind-averted side, snowdrifts several meters deep are found. One other insight from experience: when windspeeds are low to moderate, the most critical drifts are found adjacent to ridgelines; whereas when winds are strong they are found further into the slopes, at edges and often at the

lee

Dangerous drifted masses are often found in lee terrain near ridgelines. These are usually easily recognized, adjacent to completely windblown terrain on upwind slopes.

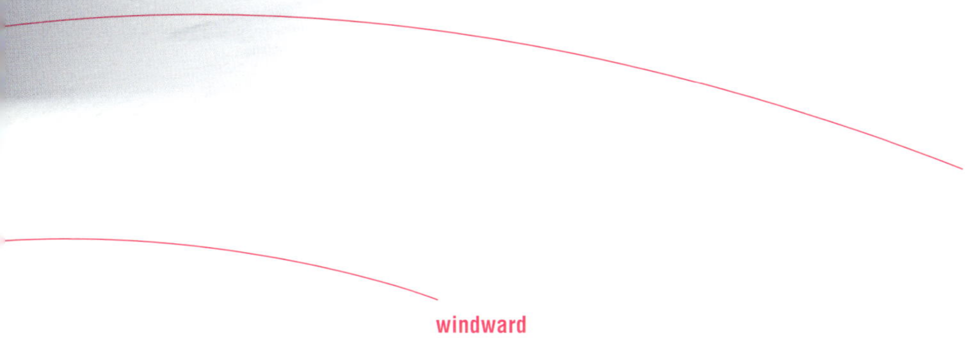

foot of cliff walls beneath rocky protruberances; and when winds reach storm strength, the dangers are far reaching and widespread throughout the landscape.

☐ New snow as weak layer

Without weak layers there are no slab avalanches. The structure of weak layers is highly varied. Among the "classic" weak layers are kinetically metamorphosed layers of faceted snow crystals, depth hoar and surface hoar. These are characterized by bonding very slowly with the surrounding layers, once they have been buried. Loose, new snow buried by snowdrifts which becomes a weak layer is completely different. This can occur either during or immediately following snowfall. If the former, there is no wind blowing when the snowfall sets in. Winds arise only during the course of the snowfall. If the latter, the entire new snow is first deposited loose and only afterwards (up to 2 days afterwards) does wind arise and transport the fresh fallen snow to leeward slopes, where it is deposited on top of loose, new snow. The potential bed surface for avalanches is located immediately at the borderline between the drifted snow and the loose, new snow. Both layers bond relatively quickly (usually in one day). ▌

avalanche hohe munde.

Avalanche. The Hohe Munde has evolved from a spottily enjoyed springtime destination of back-country skiers to a highly frequented "all-season" touring destination. Just before Christmas, two German skiers are on their way to the summit. They trigger a slab avalanche in the upper third in steep terrain. Both skiers are swept along. One of them comes to a halt, injured, on the surface of the plummeting snow, the other is swept further through rocky terrain and suffers fatal injuries.

Short analysis. A brief intermediate high promises cold morning temperatures and generally cloud-less skies. The landscape is wintery as the result of 30 cm of recently fallen new snow. The old snowpack in the area of the accident is stable. The danger stems from recently formed snowdrift accumulations which are found only above the treeline. Below the treeline, superb powder snow deposited on a hardened old snowpack can be enjoyed. Above the treeline, however, northerly winds had previously been blowing which grew stronger with ascending altitude. In corresponding manner, the spread of snowdrift accumulations also became more far reaching with ascending altitude. The situation was also particularly treacherous because towards the end of the period of snowfall, winds slackened off and as a result, the snowdrifts themselves were buried by loose fresh snow. It is difficult to assess this danger. Nonetheless, also on the day of the accident it was windy at high altitude, and additional, fresh, trigger-sensitive drifts were accumulating. Mountain rescue squad members at the scene of the accident reported that cracks and fissures opened in the snowpack when it was tread upon.

Where Hohe Munde / western Northern Alps / 2350 m / SE-slope / 40°

Where 2 persons involved / 1 fatality, 1 injured person

When 20. 12. 2012, 12:00 midday

Avalanche slab avalanche (dry) / L 650 m / W 50 m / fracture 0.3 m

Regional danger level 2 (moderate)

Aval. bulletin headline Heightened proneness to snowpack triggering esp. above 2200 m on W/N/E facing slopes

Avalanche problem wind-drifted snow

avalanche lafatscher joch.

Avalanche. Two Bavarian backcountry skiers plan to traverse the Karwendel Massif. They start out in Giessenbach near Scharnitz on 4 March. When they do not report to their families on 8 March as planned, the alarm is sounded. During a search flight on 9 March at midday, an avalanche is discovered below Lafatscher Joch in which a lifeless body is visible. The second person is not found until mid-May in the course of an elaborate search.

Short analysis. On 5-6 March, Tirol lies in the path of a powerful SW foehn air current. A storm-strength wind is blowing in the mountains, particularly in the classic foehn lanes, including on Lafatscher Joch, where the southerly wind is channeled intensively. Visibility is adequate, with some dispersed cloud. Prior to the foehn storm, favorable conditions predominated, with slopes full of terrific powder snow. These are transformed within the shortest imaginable time to dangerous traps full of snowdrifts. An entry in the log book at a refuge records that both ski tourers do indeed recognize to a certain extent the danger signs created by foehn winds. Nevertheless, they still attempt to cross the transition zone towards the Hallerangerhaus refuge through terrain which in places is very steep. They take the summer route and trigger the avalanche in moderately steep terrain over a snowpack surface which has been hardened by wind. It is a hard and very brittle slab which had formed just prior to their crossing. An important principle in practice is that a hardened snowpack surface does not always ensure safety. It is also important to remember that hardened slabs are particularly easy to trigger during a period of cold and during or immediately following storm winds.

Where Lafatscher Joch / western Northern Alps / 1950 m / N-slope / 33°

Who 2 persons involved / 2 fatalities

When 6. 3. 2013, time unknown

Avalanche slab avalanche (dry) / L 230 m / W 70 m / fracture 0.3–1.5 m / burial 3 days and 2 months

Regional danger level 2 (moderate)

Aval. bulletin headline Beware snowdrift accumulations in high alpine regions

Avalance problem wind-drifted snow

danger pattern.7
shallow snow next to deep snow

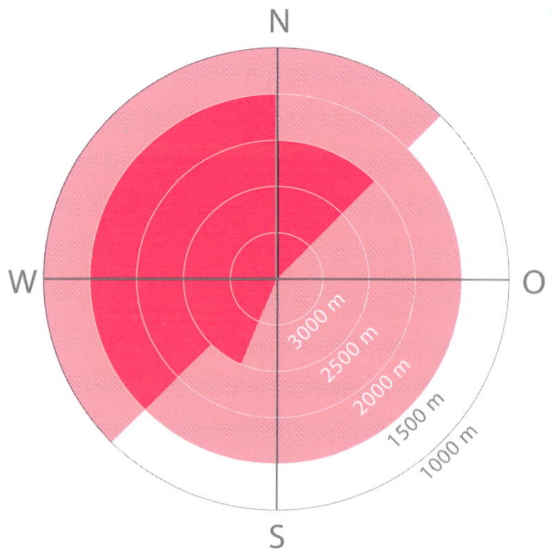

The graph illustrates winters of heavy snowfall with pronounced NW weather.

Areas where the snow is shallow usually have less favorable snowpack layering than where the snow is deep. This is because of amplified transformation processes which go on inside a shallow snowpack. Furthermore, slab avalanches are easier to trigger where the snow is shallow because the weak layers are not as deeply embedded in the snowpack, thus are more liable to trigger through additional loading. For that reason, avalanche triggerings are often observed in transition zones from shallow snow to deep snow. This occurs frequently near terrain edges, and repeatedly on broad ridges.

avalanche silleskogel.

This example illustrates one of the very rare accidents out of which it is not possible to perceive its trigger mechanism in the conditions previous to its triggering. Only directly at the avalanche fracture can the riddle be solved and the true cause be revealed.

□ Circumstances of the accident

Two local backcountry ski tourers ski singly onto the already somewhat tracked summit slope of the Silleskogel in the Zillertal Alps. While the first person reaches the foot of the slope without any problem and waits there, the second person swiftly triggers a medium sized slab avalanche as soon as he skis onto the slope. The avalanche unleashes over the ascent route as well as the spot where the first person stands waiting. However, he manages to flee the avalanche path in time and exit from the danger zone. The skier who is swept along with the plummeting snow masses is able to inflate his airbag and is buried up to his head, with only minor injuries and can free himself with the help of the other skier. Fortunately no one else was inside the rather wide swath of the avalanche on the slope.

□ Analysis

Weather and snowpack. Variable, unseasonably cold weather condtions accompanied by snowfall; mostly light winds, intensifying with ascending altitude; rather a lot of sunshine characterizes the

Where Silleskogel / Zillertal Alps / 2360 m / N-slope / 40°

Who 2 persons involved, one injured person

When 1. 4. 2013, 1:00 pm

Avalanche slab avalanche (dry) / L 300 m / W 250 m / fracture 0.4 m / burial 10 min.

Regional danger level 3 (considerable)

Aval. bulletin headline Considerable high alpine avalanche danger regionally – Beware fresh snowdrifts!

Avalanche problem new snow / persistent weak layers

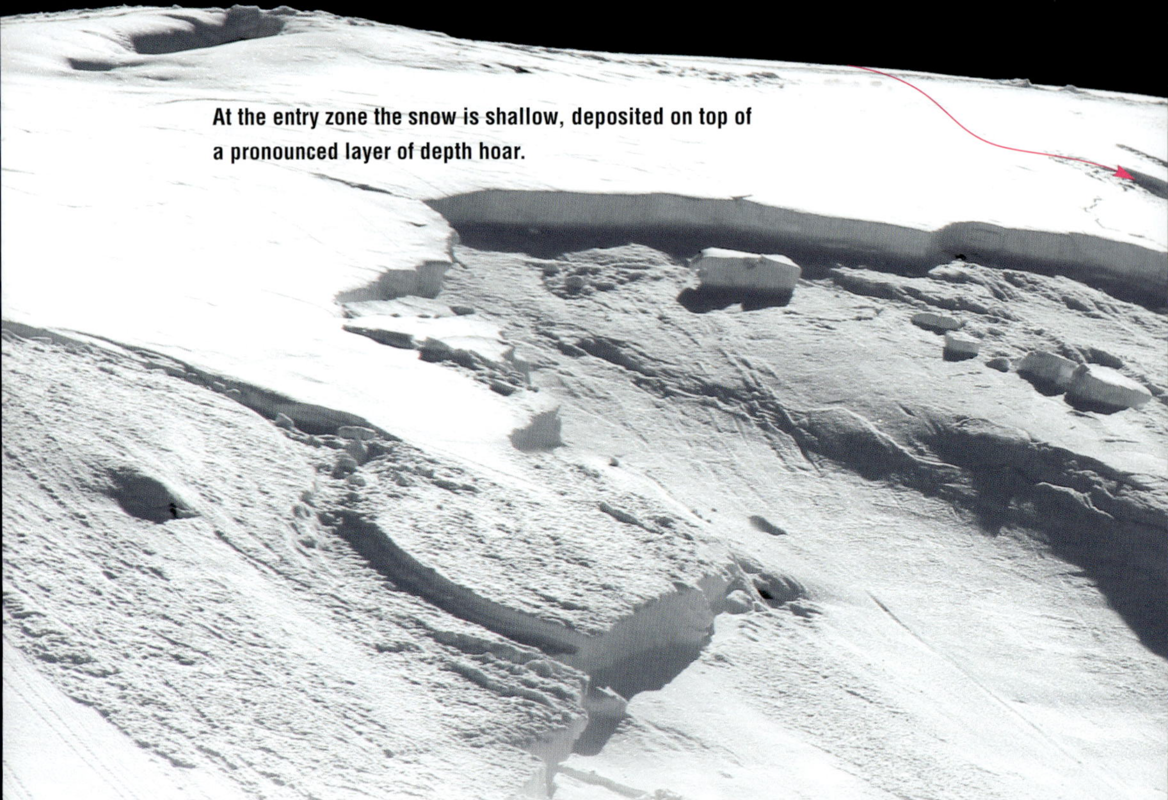

At the entry zone the snow is shallow, deposited on top of a pronounced layer of depth hoar.

prior Easter holidays. The air is intermittently very dry. One frequently encounters absolutely fabulous Canada-type powder snow, most of it in western regions, where in some areas the fresh fallen snow increases to significant depths with ascending altitude. In the area of the accident the snow is 30-40 cm deep. The snowpack surface becomes more and more irregular with each passing day. On sunny slopes, breakable melt-freeze crusts form to an increasing extent; on shady slopes powder snow can be enjoyed but is increasingly wind-impacted at higher altitudes.

Wide ranging snowpack investigations showed that the old snowpack was by and large quite compact, that is, so low in tensions that avalanche prone locations could practically not be found. The easily assessed main problem lay in fresh snowdrifts at high altitude.

The following three factors were decisive for the avalanche:

- Due to a peculiar constellation of weather conditions, a thin, faceted layer began to form starting on 27 January. The layer was recognizable. However, all stability tests showed it to be well bonded without any tendency towards fracture propagation.
- The recently fallen fresh snow had settled in the interim and could thus absorb tensions.
- What mattered most was the discovery which we made at the entry track at the point of fracture. There we found an area where snow was shallow and massive depth hoar evident at ground level. In some places the snowpack lying on top of it was well bonded, at other places, not. The backcountry

Impressions during the ascent

skier must have touched a hot spot and triggered a slab in his very first turn on the slope. Only through the additional loading on top of this small slab did the thin, faceted layer fracture. This explains the surprising magnitude of the avalanche.

Avalanches. There were no visible warning signs in the form of freshly triggered avalanches. On sunny slopes, loose-snow avalanches can be observed in isolated cases. They are harmless, however.

Terrain. The avalanche was triggered in extremely steep, shady terrain adjacent to a ridgeline. What was decisive was that during the winter, the snow remained very shallow, due to wind influence. This stimulated kinetic (faceting) metamorphism of the snow at ground level.

Bad luck played a part in this avalanche. Nevertheless, if one can possibly learn something from it, then the insight that in spite of a generally well-structured snowpack, slab avalanches can still trigger in very steep terrain, particularly where the snow is shallow, not deep. Thus, one should tend to avoid those areas where snow is shallow and which, incidentally, are easy to recognize in the entry zones to a slope. ▌

dp.7 recognizing it

When there has been heavy snowfall, places where the snow is shallow are usually easily recognizable. It is often terrain adjacent to ridgelines. Quite regularly such places also occur on slopes exposed to heavy wind impact. This danger pattern is found very often in Tirol, as a result of its frequent NW weather conditions.

One clear indicator of places where the snow is shallow is supplied by rocks which jut out through the uppermost surface on one aspect, whereas in other aspects the rocks are buried deep in snow.

background knowledge: shallow snow next to deep snow.

dp.7 definition

In winters where snowfall has been heavy, the transition zones from deep to shallow snow are critical. In case there is a trigger-sensitive weak layer inside the snowpack, it can trigger more easily in places where the snowpack is shallow than where it is deep.

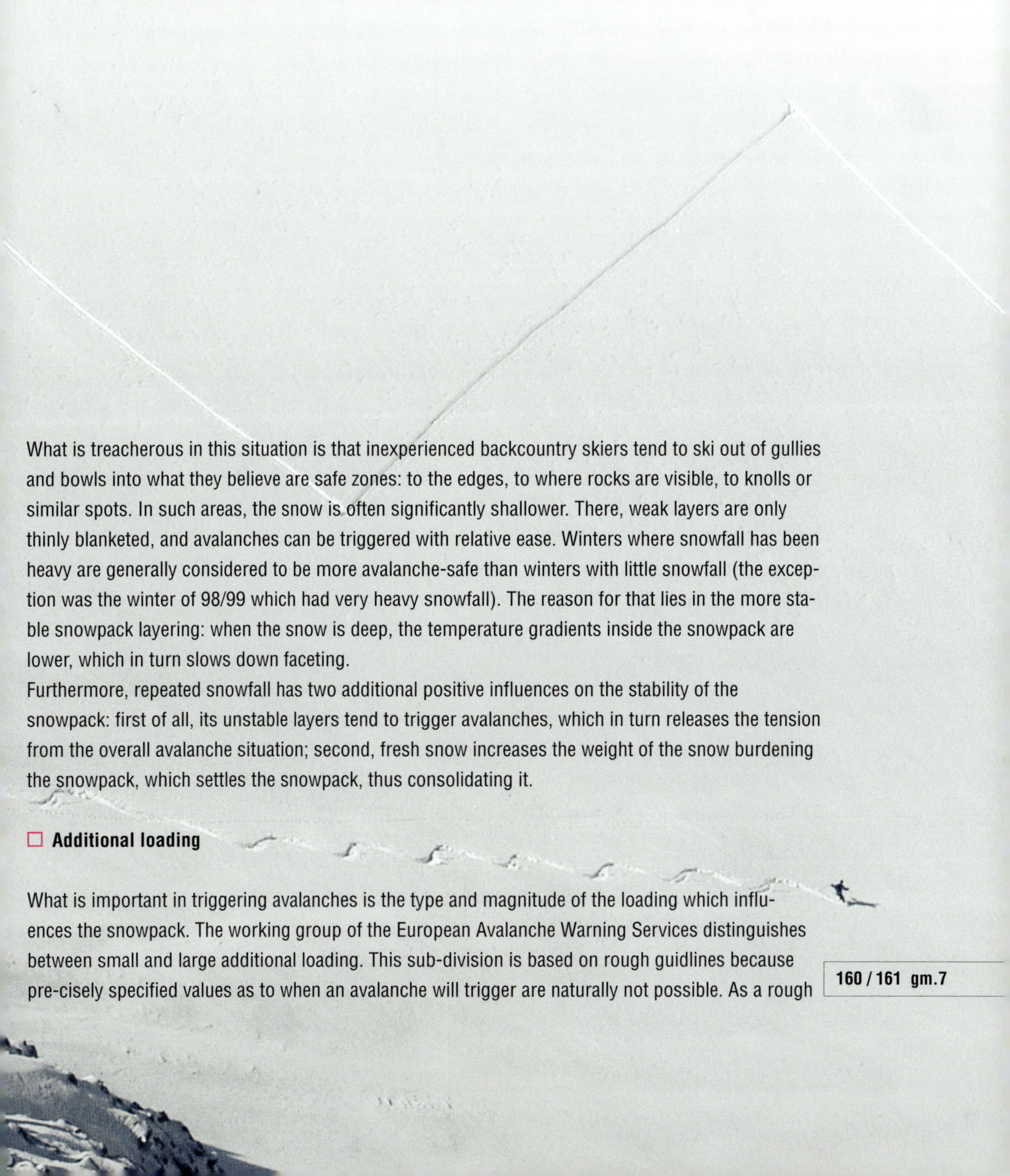

What is treacherous in this situation is that inexperienced backcountry skiers tend to ski out of gullies and bowls into what they believe are safe zones: to the edges, to where rocks are visible, to knolls or similar spots. In such areas, the snow is often significantly shallower. There, weak layers are only thinly blanketed, and avalanches can be triggered with relative ease. Winters where snowfall has been heavy are generally considered to be more avalanche-safe than winters with little snowfall (the exception was the winter of 98/99 which had very heavy snowfall). The reason for that lies in the more stable snowpack layering: when the snow is deep, the temperature gradients inside the snowpack are lower, which in turn slows down faceting.
Furthermore, repeated snowfall has two additional positive influences on the stability of the snowpack: first of all, its unstable layers tend to trigger avalanches, which in turn releases the tension from the overall avalanche situation; second, fresh snow increases the weight of the snow burdening the snowpack, which settles the snowpack, thus consolidating it.

☐ Additional loading

What is important in triggering avalanches is the type and magnitude of the loading which influences the snowpack. The working group of the European Avalanche Warning Services distinguishes between small and large additional loading. This sub-division is based on rough guidlines because pre-cisely specified values as to when an avalanche will trigger are naturally not possible. As a rough

When a weak layer is thinly blanketed (A) an avalanche can be triggered by the weight of one skier. When the weak layer is buried by enough snow (B) that is no longer possible.

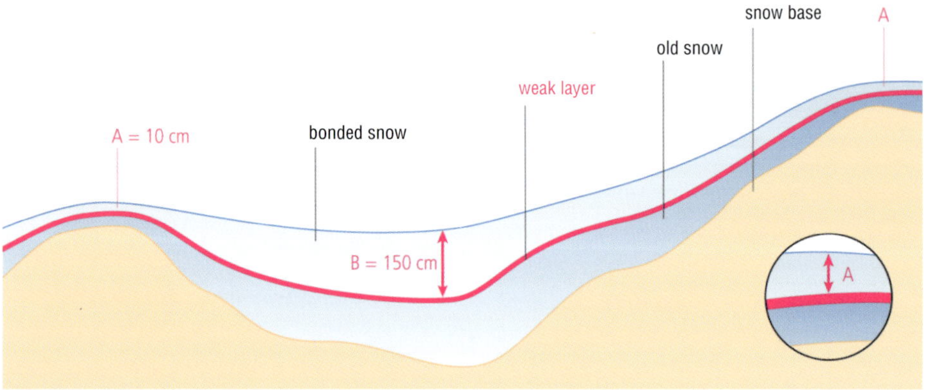

rule of thumb: when on skis, snowboards or snowshoes the loading is reduced (less pressure due to larger surface, possible weak layers are not as easily reached and disturbed); that also applies to maintaining distances during ascent, and skiing individually down a slope for the descent. Conversely, it means one single mountaineer climbing without skis disturbs the snowpack more than a whole group of backcountry skiers maintaining distances. And a skier who takes a fall loads the snowpack more than several people standing still.

A single skier or snowboarder can impact, i.e. disturb the snow cover only to a certain limited depth, due to his weight. In soft snow, the depth is about 80-100 cm; in well bonded snow (snowdrifts) the disturbance can barely reach 50 cm. Furthermore, when the snow on the uppermost surface is hard, the disturbance of a skier's weight is distributed over a larger surface, thereby preventing the load from penetrating to more deeply embedded layers. If the upper surface is soft, the disturbance penetrates more deeply, but has lesser impact in terms of surface area. That means, if there is sufficient bonded snow on top of a weak layer, e.g. in the middle of a bowl, one individual skier can hardly disturb this weak layer. But if at the edge of a bowl or in a transition zone to a windblown knoll there is only shallow snow deposited on top of a weak layer, the impact can be disturbed and subsequently, an avalanche can be triggered. ▌

Taking a fall is considered additional loading of approximately 10 times the weight of the skier.

large additional loading

- two or more skiers, boarders, etc. not maintaining distances
- ski slope motor vehicle, avalanche explosion
- one sole hiker/mountaineer

little additional loading

- sole skier or boarder, making gentle turns, not falling
- group maintaining distances (minimum 10 m)
- snowshoe hiker

avalanche valluga.

Avalanche. The north-face descent down the Valluga is known for its extreme steepness and exposed terrain. This is free skiing terrain which must be understood and evaluated by each individual. Access to this descent via a small cabin cablecar is permitted only to guided groups. On the day of the accident a mountain guide selects the descent towards Zürs with his six guests. The guide traverses a slope then waits with one guest at a meeting spot, while four guests descend and one waits at a previous meeting spot on a small saddle. The four persons descending on skis trigger a slab avalanche which sweeps all of them away with the plummeting snow masses. One of them comes to a stop above a rock fall with bruises and contusions; the other three are swept over rocky terrain but also not buried. However, in one case grave injuries are suffered, in two cases fatal injuries.

Short analysis. The winter of 2014/15 is known for its pronounced persistent weak layer problem with deep persistent weak layers from early winter. The accident slope has hardly any of this, however, since the relevant meteorological influences did not exist at this altitude. In order to find the cause of the avalanche, snow profiles and stability tests were conducted at various spots on the slope, including profiles which proved to have extremely unfavorable, as well as extremely favorable layering. It turns out that the cause of the avalanche is to be found in the small-surface area but highly trigger-sensitive old snowpack with shallow snow in places. There, during the winter, layers of pronounced faceted snow crystals developed beneath embedded crusts, alongside depth hoar. Thus, it was in one of the classic hotspots where this avalanche was triggered.

Where Valluga, Arlberg – Ausserfern / 2790 m / N-slope / 40°

Who 7 persons involved / 2 fatalities and 2 injured persons

When 19. 1. 2015

Avalanche slab avalanche (dry) / L 800 m / W 55 m / fracture 0.2–1.9 m

Regional danger level 2 (moderate)

Aval. bulletin headline Ongoing treacherous avalanche situation in some places above 2100 m – Rainfall: beware glide-snow avalanches!

Avalanche problem persistent weak layers

avalanche malgrubenspitze.

Avalanche. On a beautiful spring day two brothers climb from the parking lot of Axamer Lizum over very steep, shady terrain up to the Malgrubenscharte. Then, maintaining distances, they ski onto the slope. After only a few turns, one of them triggers a slab in a spot where the snow is shallow. Both persons are swept along. One of them just manages to escape, whereas the other is swept further and is completely buried. Since the buried skier, as it later proves, had an avalanche transceiver but it was not activated, the search proves difficult. He can be pinpointed and freed from the snow masses only after two hours of intense search by the rescue squad. He dies the same day in the hospital.

Short analysis. By the time of the accident, two periods of cold during the winter have left behind striking effects, especially in shady terrain. There, near ground level, one finds weak layers from early winter which, however, are not very prone to triggering. This is because the weak layers have consolidated somewhat in the interim; and because the influence of springtime warmth on the accident slope had not yet caused water seepage into the snowpack which would have led to loss of firmness. During the accident analysis it becomes evident that the snow is far shallower in the upper third of the slope than further down, certainly a result of wind impact. This also explains why the deep persistent weak layer is far more pronounced in the upper reaches of the slope: because of intensified processes of transformation in shallower snow. ▌

Where Malgrubenscharte / northern Stubai Alps / 2380 m / N-slope / 40°

Who 2 persons involved / 1 fatality

When 22. 4. 2010, 10:15 am

Avalanche slab avalanche (dry) / L 350 m / W 40 m / fracture 0.1–0.7 m / burial 2.5 m / 2 hrs.

Regional danger level 2 (moderate)

Aval. bulletin headline Favorable conditions in the morning – daytime increase in avalanche danger

Avalanche problem persistent weak layers

danger pattern.8
buried surface hoar

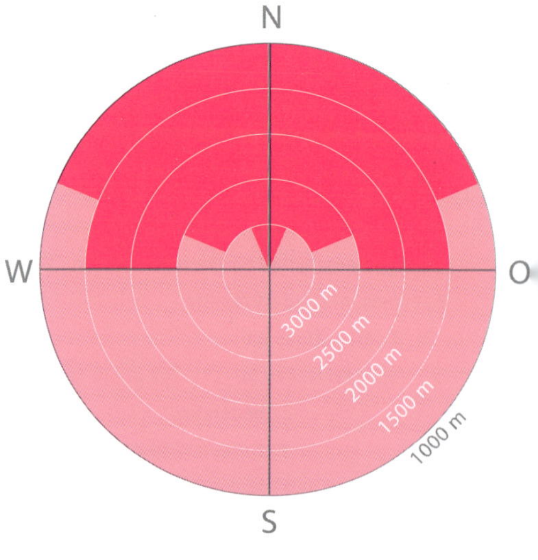

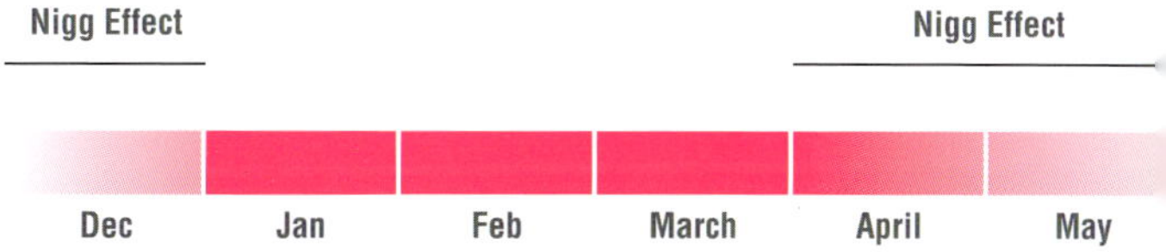

Surface hoar numbers among the most beautiful types of snow in existence. In and of itself, it bears no danger potential. That potential unfolds only when it is buried by fresh, bonded layers of snow. Then it becomes hazardous. In fact, it is considered one of the most critical weak layers in all of snow and avalanche science.

avalanche galtenberg.

This is a personal tale by one of the authors. Following an active period with many avalanches, some of which caused accidents involving people and fatalties, the task was to bring a backcountry touring group from an Alpine Club safely up the Galtenberg in the Kitzbühel Alps. Had the appropriate background knowledge been lacking, the likelihood of stumbling into one of the many already armed traps would have been great.

☐ Avalanche

On the Galtenberg summit, we savor the magnificent views and are happy about our successful attainment of the highest point of the peak. We know that massive surface hoar has recently formed. This awakens the curiosity of the author to cast a glance into the snowpack. Maybe it might even be possible to trigger a slab avalanche from a safe spot into an empty bowl of terrain? We undertake a thorough search for a suitable location. Ultimately, all that is necessary is to carefully tap a foot from a safe crest onto the steep E-NE facing slope.

We immediately discover fissures. Very softly then, a slab is released which roars down towards the valley over partially rocky summit flanks. Immediately the author reports the avalanche to be a "negative avalanche" at the reporting station of the avalanche service so that no unnecessary rescue effort is mounted. The avalanche impresses all the participants of the tour so much that during the descent

Where Galtenberg / Kitzbühel Alps / 2420 m / E-NE-slope / 45°

Who 1 person involved

When 13. 3. 2005, 2:30 pm

Avalanche slab avalanche (dry) / L 400 m / W 90 m / fracture 0.5 m

Regional danger level 3 (considerable)

Aval. bulletin headline Considerable avalanche danger regionally

Avalanche problem new snow

there is not one word of discussion about whether to perhaps take other, steeper slopes of descent. On our selfsame "defensive" ascent route, we safely reach the valley floor again.

□ Analysis

Weather and snowpack. The winter of 2004/05 is characterized by wintery temperatures lasting well into March. In fact, the coldest days of the entire season are registered in March. At the beginning of the month, high pressure weather conditions still prevail. Only in the second week does heavy snowfall in the northern barrier cloud regions set in. It is accompanied by strong NW winds. These are not auspicious signs for the snowpack, where there still exists a trigger-sensitive fundament of depth hoar from early winter. Inside the old snowpack, a number of loose intermediate layers have formed, so that by the end of February and the beginning of March, ideal conditions prevail for the formation of surface hoar. Snowfall and wind then give rise to wide ranging snowdrift accumulations which are highly prone to triggering. Settling noises and glide cracks forming on the surface accompany the backcountry skiers every step of the way.

dp.8 recognizing it

Without prior knowledge about the formation of this weak layer, buried surface hoar cannot be recognized. Important prior information about buried surface hoar is obtained either through one's own observations before the new snow blankets it; through independent snow analysis; or by reading the avalanche bulletin. An additional source of information is supplied by the prior week's weather: surface hoar often forms during a cool period of fine weather.

dp.8 recognizing it

A good reference point in open terrain is supplied by naturally triggered avalanches with relatively shallow fracture depths. Small avalanches are especially instructive, e.g. interspersed between avalanche-braking barriers or on steep hillsides.

Avalanches. We are in one of the most avalanche-active periods of the entire winter. Between 10 and 12 March avalanches are reported almost hourly, many of which bury people in their path. There are also numerous naturally triggered avalanches which unleash not only during snowfall, but also as a consequence of significant rises in temperature and increasingly intensive solar radiation.

Terrain. The avalanche on the Galtenberg is triggered in a very steep spot adjacent to a ridgeline. It is interesting that the subsequent avalanche path melds into terrain with only a very slight steepness gradient. This is due to the pronounced layer of surface hoar, which shows that the snowdrifts are not bonded to the snowpack beneath them. ▌

background knowledge: buried surface hoar.

dp.8 definition

Buried surface hoar exists when moisture in the air settles on top of a cold snowpack surface and forms transparent, platelet-shaped ice crystals which are subsequently covered by new snow. The moisture necessary to forming surface hoar can also rise up from inside the snowpack.

Outgoing radiation of the snowpack is important in forming surface hoar. The snowpack must be colder than the air.

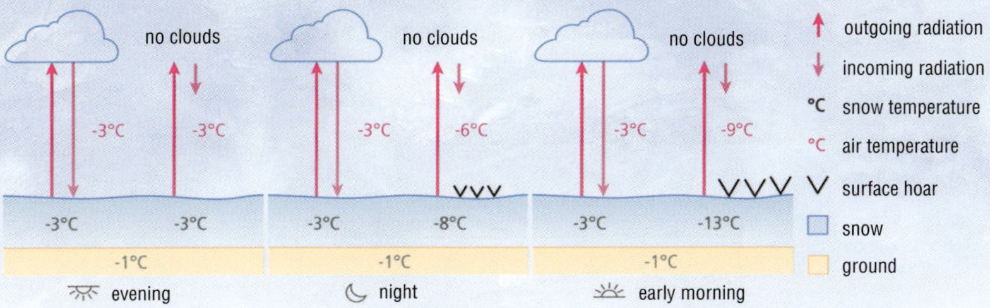

☐ Formation

Hoar does not fall from the sky, as does other precipitation, it is formed on the snowpack surface. For surface hoar to form, warmer, moist air accumulates on the snowpack surface. If the surface hoar waxes large, e.g. if it continues to grow over the course of several nights in similar conditions, and is then buried by a large amount of new snow, the buried surface hoar becomes the perfect bed surface for avalanches.

If, following cold nights of clear skies, particularly in spring or autumn, dew forms on meadows or other hypothermic surface areas, one speaks of condensation. If the dew point (= temperature to which air must cool for there to be condensation) is below 0°C, hoarfrost forms. This process, in which a direct transfer takes place from gas (water vapor) to solid (hoarfrost crystals) without passing through a liquid state is called deposition in meteorological science.

☐ Outgoing radiation

Outgoing radiation of the snowpack is important for the formation of surface hoar. The snowpack must be colder than the air. Outgoing radiation is the emitting of warmth by the snowpack surface into the atmosphere. When skies are clear, the snowpack surface cools off rapidly to below the air

Danger pattern 8 is especially easy to observe on the banks of streams.

temperature (from a few to about 20 degrees). Nights which are clear, and thus also cold, lead to a starkly hypothermic snowpack surface In the early morning hours, surface temperatures far below 0°C are often measured. The likelihood of surface hoar forming is quite high when skies are clear. When skies are overcast, the outgoing radiation is less and thus, there is less likelihood of hoar forming.

□ Microclimate at stream beds

The same thing comes about when air, for example near stream beds, is moist enough and the snowpack surface is cold enough: the surface also forms ice crystals by depositing moisture from the air, in other words, surface hoar.

A few hours later, the surface hoar is buried by fresh fallen snow and shortly thereafter avalanches unleash along the entire banks of the stream. If surface hoar is buried by new snow and there is wind impact at the same time, a critical avalanche situation immediately is born. Surface hoar forms an extremely weak, loose layer of unbonded crystals inside the snowpack. If these are covered by a layer of bonded snow, the additional loading can very frequently not be borne. That is when even small amounts of fresh snow can naturally trigger slab avalanches.

weak layer
bonded new snow
bonded old snow
snow base

Relatively warm, moist air moves over a cold snowpack surface The surface hoar in the process of forming is visible in the foreground.

☐ The Nigg Effect

The so-called Nigg Effect (named after a Swiss mountain guide) is a special form of surface hoar. It occurs when relatively warm, moist air moves over a crest or ridge. If on the rear flank of the mountain the snowpack surface is significantly cooler, due to shade or outgoing radiation, the moisture in the air is re-sublimated into surface crystals of ice. This effect is normally limited to the immediate area of the crest. A hundred meters of altitude lower down, it is not to be found.

The Nigg Effect is difficult to evaluate even for professionals, particularly if the surface hoar has already been buried by new snow. Then, an extraordinarily treacherous situation occurs. The only thing that helps in such a situation is lots of experience, precise weather and landscape observations, and when in doubt, a snow profile of several layers of the snowpack in steep, shady terrain adjacent to ridgelines in order to uncover this highly trigger-sensitive layer. The Nigg Effect is observed quite often at high altitudes in early winter and in springtime.

Just such a treacherous situation proves fateful to a group of four local skiers on a backcountry venture on 6 April 2007 on their descent from the southern Löcherferner glacier. Loose powder snow blankets a thin layer of snowdrift on a very steep slope near a ridgeline which was deposited on top of a small area of surface hoar. ▌

dp.8 recognizing it

The Nigg Effect is difficult to evaluate even for professionals, particularly if the surface hoar has already been buried by new snow. Then, an extraordinarily treacherous situation occurs. The only thing that helps in such a situation is lots of experience, precise weather and landscape observations, and when in doubt, a snow profile of several layers of the snowpack in steep, ridgeline, shady terrain in order to uncover this highly trigger-sensitive layer. The Nigg Effect is observed quite often at high altitudes in early winter and in springtime.

avalanche wannenkopf.

Avalanche. In early afternoon a group of four skiers decides to tackle the eastern flank of the Wannenkopf in the Allgau Alps near the Bolgental. Just 200 m below the summit they have to ski a very steep, short slope. They decide to do it individually. The first two persons easily reach the meeting place agreed upon, on the opposite side of the slope. The other two persons then traverse the slope together and trigger a small slab avalanche which sweeps both of them along with it and buries them totally. The two are found within a very short amount of time and freed from about 1 meter of snow. One of them suffers only minor injuries. The other dies in the hospital that same day.

Short analysis. This example was supplied by our colleagues in Bavaria. We took their story expressly for this danger pattern since surface hoar occurs far more often in Bavaria than in Tirol, due to fog occurring more frequently there. Surface hoar forms especially often at the upper altitude limits of fogbanks. Furthermore, surface hoar persists for particularly long periods in shady, sparsely wooded regions because of the lack of wind in such terrain. In this specific accident, fog does not play a major role. The cause lay in the unusually cold winter, which provides near-perfect prerequisites for surface hoar. Beginning on 9 March, nearly 50 cm of fresh snow fell in several bouts, accompanied by storm-strength winds. On 11 March, the air temperature suddenly surged noticeably. It was this combination which unleashed numerous naturally triggered avalanches in the immediate vicinity. ▌

spot of burial

Where Wannenkopf / Allgau Alps – Bavaria / 1710 m / E-slope / 37°

Who 4 persons involved / 1 fatality

When 11. 3. 2005, 12:00 p.m.

Avalanche slab avalanche (dry) / L 80 m / W 30 m / fracture 0.3 m / burial 1 meter

Regional danger level 4 (high)

Avalanche problem new snow

avalanche wiesejaggl.

Avalanche. A snowboarder decides to descend the extremely steep summit flank of the Wiesejaggl together with three friends. They approach the entry point from a ski area via a short, steep upward sweep. The British boarder is the first to enter the extremely steep slope. After only a few turns he triggers a slab avalanche which initially is quite small, but subsequently sweeps away more deeply embedded layers of the snowpack, grows to large size and separates at the foot of the slope into two different arms. That is where he is buried. His friends send an alarm to the mountain rescue squad and then descend to the spot where the avalanche separates into two arms. They begin their search in the wrong arm. Soon, however, they pinpoint the location of their buried friend and dig him out with the help of the mountain rescue squad. Their efforts to resuscitate him fail.

Short analysis. Starting at the end of April, variable weather conditions prevail, a mixture of snowfall, clouds and sunshine, with light-to-moderate strength winds which often shift direction. Temperatures correspond to the juncture of the season. Warm, moist air masses move over the crest of the Wiesejaggl during a foehn current at the end of April. They leave behind a thin ribbon of surface hoar extending to about 25 meters below the crest. It can also be seen after the accident on-site that this layer of buried surface hoar is 2 cm thick just below the crest, then gets continually thinner with increasing distance from the crest. The buried surface hoar is extremely prone to triggering at the moment the accident occurred. This weak layer could only have been recognized if one had been roped up and dug into the snowpack. ❚

Where Wiesejaggl / southern Ötztal Alps / 3120 m / N-slope / 45°

Who 4 persons involved / 1 fatality

When 5. 5. 2009, 2:45 p.m.

Avalanche slab avalanche (dry) / L 350 m / W 250 m / fracture 0.2–1.5 m / burial 1.5 m

Regional danger level 2 (moderate)

Aval. bulletin headline Moderate danger widespread – Beware small snowdrifts in high alpine regions

Avalanche problem new snow / persistent weak layers

danger pattern.9
buried graupel

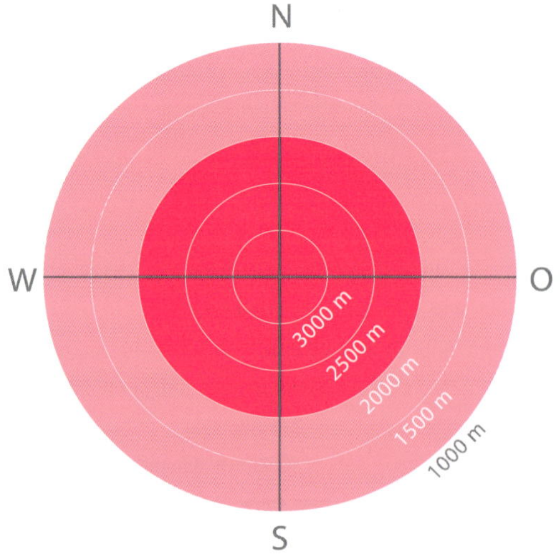

Jan | Feb | March | April | May

A weak layer inside a snowpack is often compared with ball bearings. This image is actually suitable only for graupel: a spherical, granular form of precipitation which is often deposited during showers such as thunderstorms during springtime. It is easy to grasp that snowdrifts accumulating atop of a layer of graupel can only be poorly bonded with it, so that the danger of avalanches increases. Graupel is frequently spread over very small surface areas. Without examining the snowpack itself it is difficult to spot, even for professionals. This is an extremely treacherous phenomenon. Fortunately, the problems it creates are short-lived.

avalanche granatenkogel.

This avalanche accident below the Granatenkogel in the southern Ötztal Alps was formidable in several ways. First of all because of the magnitude of the avalanche and the reports of observers that the whole mountain seemed to quake when it released. Then, because of the similarity to a particularly tragic accident on the nearby Schalfkogel nearly five years later on 2 May 2009. Fortunately in this avalanche, people got off lightly.

□ Avalanche

A group of six highly qualified backcountry ski tourers climb on a warm spring day from Obergurgl over the Ferwalltal to the Granatenkogel, 3318 m. All reach the summit. During the descent over the crest, a slab avalanche triggers naturally just below their track and grows to enormous size. The group is startled but safe. They are treated to a new experience, a fascinating if horrifying drama of nature. In the runout zone of the avalanche, no people are caught. Thus, the avalanche caused no damage.

□ Analysis

Weather and snowpack. It is the finale of the second-warmest winter since the beginning of measurement-taking, a winter which remains in memory in two ways. In northern regions there

Where Granatenkogel / southern Ötztal Alps / 3300 m / N-slope / 45°

Who 6 persons involved

When 1. 5. 2014, 10:15 a.m.

Avalanche slab avalanche (moist) / L 700 m / W 300 m / fracture 0.2–2

Regional danger level 3 (considerable)

Avalanche problem new snow

dp.9 recognizing it

Buried graupel (unless you dig into the snowpack) is one of the most difficult avalanche-relevant factors to recognize. Basic meteorological knowledge can supply a indicator of heightened or threatening danger. Graupel is deposited during precipitation such as thunderstorms.

is too little snow; in southern regions, too much. April unfolds full of quick changes and highly varied conditions, as is its wont. This is also the case in the days prior to the accident when cloudy, moist and warm air masses normal for this juncture of the season reach Tirol. The sun is also shining, which intensifies diffuse light conditions. The quite stable old snowpack becomes more moist, more prone to triggering with each passing day. This is also because the water which has seeped into the snowpack for the first time this winter reaches a faceted layer of snow near ground level and the snowpack begins to weaken. But what is decisive are the showers of graupel at the end of April. A relatively thick layer of graupel is deposited inside a layer of fresh fallen snow and thereby forms a weak layer which must not be underestimated.

At first it seems likely that primarily a slab could trigger from this layer of graupel. Only secondarily does additional loading threaten a fracture in a deep persistent weak layer which extends over wide ranging surface areas. We can assume that over the next few days, the layer of graupel would probably not have triggered. This is because of the rapid settling of the layer of fresh snow due to warm temperatures, solar radiation and the bonding with the graupel layer.

Avalanches. Three days prior to the avalanche, naturally triggered avalanches were observed after a long interim without avalanches. Usually these are wet, loose-snow releases immediately following snowfall on extremely steep slopes. However, increasingly frequent medium-sized,

dp.9 recognizing it

Particularly in springtime, snow depths are highly divergent. This is a clear indicator of graupel shower cells. If these graupel layers are pronounced, isolated naturally triggered avalanches are the result.

in isolated cases even large-sized, slab avalanches were also triggering naturally, particularly on W/N/E facing slopes above about 2500 m. It is not at all unusual for the impulse of a loose-snow avalanche to serve as the trigger mechanism for a slab avalanche to subsequently release. Below about 2500 m, the avalanche season was over, for the most part. Frequent water seepage down to the ground brought about a rotten snowpack and this had already melted away.

As mentioned above, the situation reminds one of the beginning of May 2009. The conditions were nearly identical, both as regards weather and as regards the snowpack layering. Six mountain climbers lose their lives near the Schalfkogel.

Terrain. The flanks of the Granatenkogel are ideal avalanche terrain, usually extremely steep. Avalanches are usually triggered in a convex, ever steeper transition from the crest to the slope. The shadiness and altitude are both prerequisites for the formation of such a huge avalanche. ▌

At the beginning of May, large-sized naturally triggered slab avalanches are observed throughout the land.

background knowledge: buried graupel.

dp.9 definition

Buried graupel forms when snow crystals freeze to form hypothermic water droplets as they fall and this layer is then covered by fresh snow. Unlike hail, which consists mainly of ice and is thus glassy, there is still air inside the graupel. Thus, graupel is less dense than hailstones, softer, and milky or white in color.

Small spheres of graupel grow through fierce vertical movements in showers for as long as it takes to finally fall to the ground from their own weight.

Variety of forms

There are three different types of graupel, distinguished by the way they come about:

▌The most important form is frost graupel. It is the typical form occurring in shower or thunderstorm clouds at temperatures below -4°C. If hypothermic water droplets meet snow or ice crystals, they attach themselves to these crystals and form spherical grains. The graupel grains are heaved about by frequently powerful upwards-downwards winds inside the shower cells and thereby continue to grow (up to about 5 mm).

▌Hoar graupel seldom falls isolated, usually it falls from the sky together with snow crystals. It is formed through the thawing and renewed freezing of snow crystals in high altitude clouds. Unlike frost graupel, the powerful upwards-downwards winds are lacking.

▌Gries (like semolina) or rounded grains are the smallest form of graupel. Their diameter is 1mm or less. Gries forms at temperatures below 0°C in layered clouds, e.g. stratus, not in showers. The grains are opaque, rounded conglomerates of snow crystals.

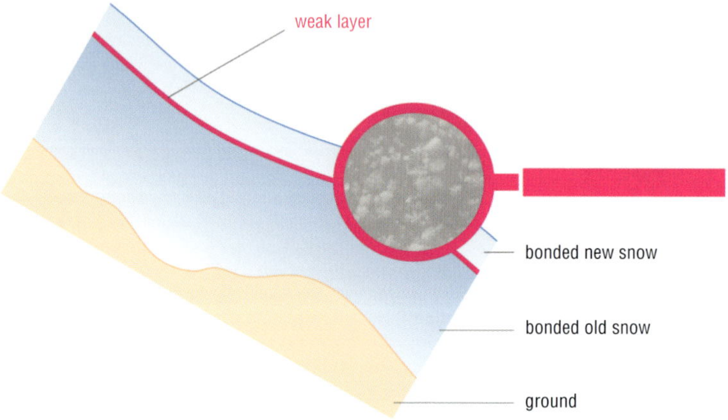

Explaining the avalanche

Regardless through what process graupel is formed, the rounded grains become an ideal bed surface for avalanches when they are buried by new snow or covered by wind-drifted snow, truly resembling a layer of ball bearings.
As to their frequency and spread in avalanches, they ordinarily cover only a small surface area. Most often, graupel is deposited in bowls and slope concavities. For that reason, area-wide, cohesive bed surfaces are the exception. The way graupel forms is interesting, similar to surface hoar. Both are beautiful in form, but difficult to recognize once they are covered with snow. Surface hoar, however, is the far more critical bed surface of the two. Absolute certainty about graupel is only obtainable from snowpack analysis, and even then, such analysis supplies a picture only at one small point of territory. It is quite possible that completely different conditions prevail 10 meters to the left or 10 meters further up.

The moment in time when graupel is formed is either during a deep-winterlike cold front or in springtime; in other words, periods when intense solar radiation and higher temperatures lead to rapid settling and stabilizing of the uppermost layers of the snowpack. Thus, the danger pattern of buried graupel tends to occur most often in spring weather conditions, just a few days after snowfall.

Graupel grains up to 2 mm in size on a snow grid.

lawine schalfkogel.

The avalanche accident below the Schalfkogel in the Ötztal Alps awoke a great deal of interest in the media. This was no coincidence, since it was the gravest avalanche accident in ten years in Tirol's outlying terrain.

☐ Circumstances of the accident

On 1 May a group of seven backcountry ski tourers comprising six Czech and one Slovak climb up to the Hochwildehaus from Obergurgl in very good weather. The refuge was closed at the time, they spend the night in the winter quarters hut. On 2 May, poor weather initially delays their start. When towards midday conditions improve somewhat, six of the group decide to climb the Schalfkogel. The seventh remains behind, choosing to recover from the efforts of the previous day.
On the way to the summit, they deviate from the normal ascent route as a result of deteriorating weather conditions. At an altitude of 3250 m, they turn back. During the descent they trigger a slab avalanche which buries all of them. Coincidentally the man running the Langtalereck refuge notices the avalanche fracture and sounds the alarm. An immediate search is not possible, due to bad weather. On the following day, a search is initiated in the early morning hours in splendid weather conditions. It is soon determined that none of the tourers, aged between 20 and 33, survived the accident.

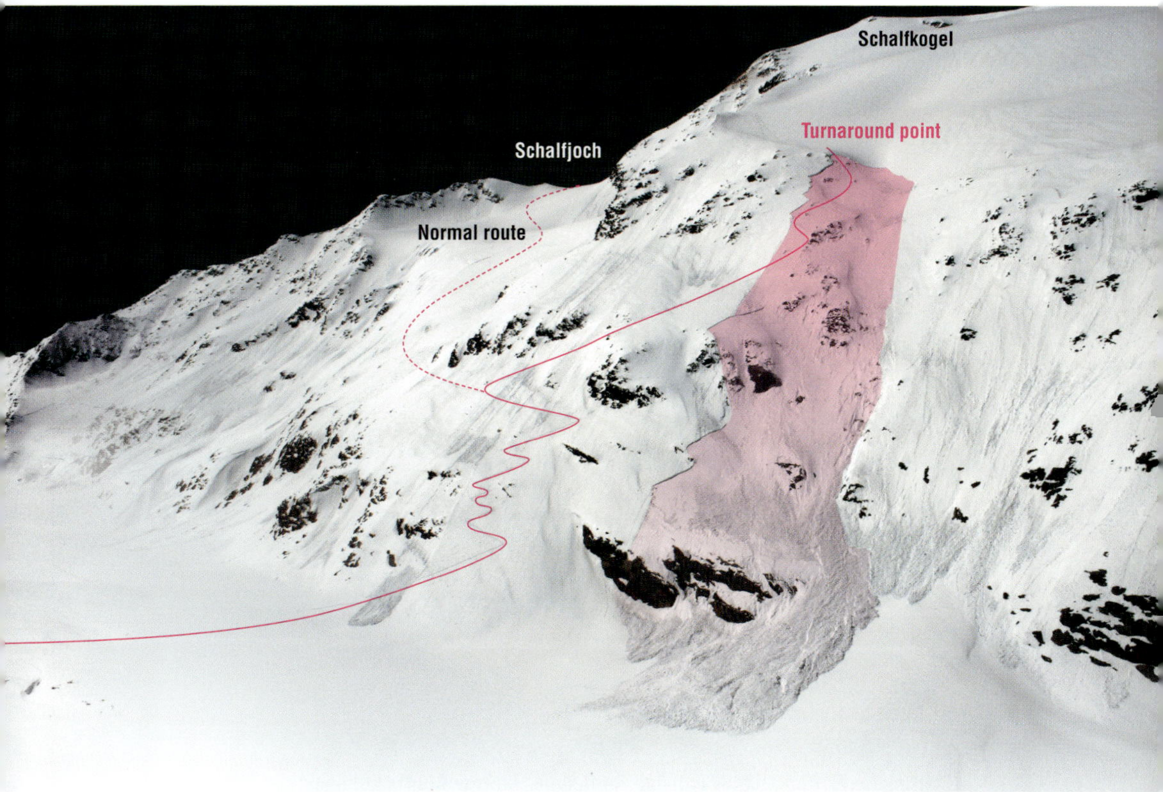

Where Schalfkogel / southern Ötztal Alps / 3180 m / E-NE-slope / 40°

Who 6 persons involved / 6 fatalities

When 2. 5. 2009, 4:00 p.m.

Avalanche slab avalanche (dry) / L 500 m / W 170 m / fracture 0.4–1 m / burial 1–2.2 m / 1 day

Regional danger level 3 (considerable)

Aval. bulletin headline Slight increase in avalanche danger due to rainfall and new snow.

Avalanche problem new snow

The backcountry skiing group spent the night in the winter room of the Hochwildehaus (Fidelitas hut).

Short analysis

Starting on 27 April, highly variable, sometimes stormy April weather sets in. Between 28 and 29 April there had been as much as 50 cm of snowfall in the accident zone, deposited on top of a hardened old snowpack surface which had formed during the previous two nights of outgoing radiation. Inside the new snow was an especially pronounced, obviously trigger-sensitive layer of graupel over a wide surface area which caused a number of naturally triggered avalanches in the rear part of the deep circular valley.

On the trail to the Hochwildehaus on 1 May, these avalanches provided a clear-cut and quite visible indication of a pronounced weak layer which should have influenced the choice of tour and route selection on 2 May. The goal of the tour is far too steep; moreover, poor visibility makes on-site assessment of dangers very difficult. ▌

A glance from the helicopter shows in awesome ways the 500-meter long avalanche path of the plummeting snow masses and the spots where the six men were buried.

danger pattern.10
springtime situation

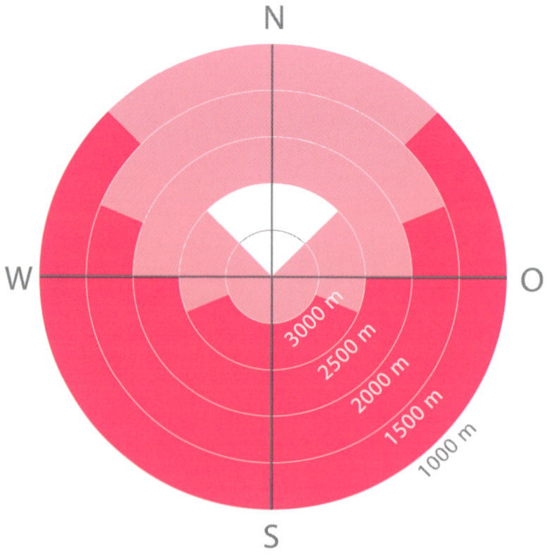

early phase (February, March)

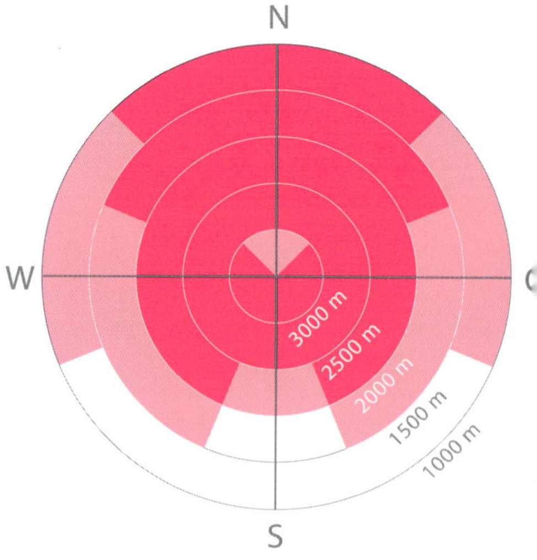

late phase (April, May)

Springtime presents a particularly delicate challenge to backcountry skiers and boarders, as well as to members of avalanche commissions. Rarely do states of "safe" and "unsafe" lie quite so near in time to each other. Thus, the span of danger levels which occur within the course of one single day is wide ranging. On the one hand, avalanche danger is never easier to evaluate than in stable corn-snow conditions; on the other, avalanches of such magnitude as in springtime are rarely registered in deep winter.

Apart from the snowpack layering, the complex interaction of air temperature, air moisture, solar radiation and wind play a major role. For backcountry skiers and boarders, high discipline in timing and flexibility of route planning are more essential than ever.

avalanche gargglerin.

Each spring there are brief periods of particularly intense avalanche activity. What is decisive is the warmth which penetrates the snowpack. Two such days were observed on the weekend of 11 and 12 April, when naturally triggered avalanches were not only registered, also a number of avalanche accidents occurred. One of them caused a fatality and awoke great interest in the media. The reason: one of the persons involved survived practically uninjured for ten hours while buried in snow.

□ Circumstances of the accident

Two backcountry ski tourers are descending the slope in Sandestal, a small side valley of the Gschnitztal. They are traversing a summertime access road at the foot of steep slopes when they are caught in the path of a naturally triggering avalanche, swept along and totally buried. The avalanche accident at first goes unobserved. Not until a person is reported missing at 6:35 pm on the day of the accident does the Alpine Police start the search in a helicopter.
A GPS-position fixing of their cell phone helps to locate the approximate spots of the persons involved. After a number of unsuccessful searches in several conical-shaped avalanche paths and checking into various summit logbooks, the right avalanche path is found in the Sandestal below the Gargglerin just before darkness falls. There the flight operator picks up signals from an avalanche transceiver.

Where Gargglerin / southern Stubai Alps / 2300 m / E-slope / 40°

Who 2 persons involved / 1 injured, 1 fatality

When 12. 4. 2015, ca. 12:00 noon

Avalanche loose-snow and slab avalanche (wet) / L 1000 m / W 170 m / fracture 0.3–2.5 m / burial 1.3 m / 10 hrs

Regional danger level 3 (considerable)

Aval. bulletin headline Critical avalanche situation at least below 2700 m, wet-snow avalanche danger!

Avalanche problem wet snow / persistent weak layers

The descent route along the road with the two burial spots.

The two persons were soon freed from the snow masses atop them. After a ten-hour period of being buried, one of them has died; the other, miraculously, is responsive although hypothermic. Both persons are found next to each other, lying beneath 1.3 meters of snow. The excessively long survival of one of the two was possible because his respiratory tracts stayed open. Moreover, an air channel had formed between two avalanche clumps which provided him with sufficient air inflow.

☐ Analysis

Weather and snowpack. During the prior days of pleasant weather, a spell of classic “hothouse” weather was promised for the weekend of 11-12 April. The air turned much more humid, the air temperature was high, the diffuse solar radiation was intense. Already on 11 April, for that reason, the snowpack surface became moist on the surface at very least up to high altitudes; below that it was thoroughly wet. Skies during the night of 11 April were overcast, preventing the snowpack from consolidating. Thus, in the early morning hours of 12 April, the snowpack was quite prone to triggering. The unbonded, wet snowpack surface was one aspect of the problem. A deep persistent weak layer from early winter, the other, since water seepage made it more and more prone to triggering. During the course of the day, weather conditions improved. Sunshine and warmth reigned in the mountains, with good visibility and intense solar radiation. This was an additional input of energy which did harm to the snowpack.

spots of burial

dp.10 recognizing it

The highly complex interaction of air temperature, moisture and solar radiation (plus wind, in some cases) makes it possible in springtime for avalanche dangers to swiftly increase and threaten menacingly within the shortest imaginable period of time, usually within 1-2 hours. What is decisive is the degree of snowpack wetness which is created by this combination of factors, leading to a rapid forfeiture of firmness in snowpack stability. This danger pattern can usually be discerned quite clearly by a study of weather station measurements, as well as by observations of weather and snowpack conditions in outlying terrain.

Avalanches. Avalanches were predestined on the day of the accident. In the avalanche bulletin, one could read the following: "Predominantly unfavorable conditions for backcountry tours prevail today. Already in the early morning hours, avalanche danger below about 2500 m is considerable; above that altitude, danger is moderate and will increase to considerable up to about 2800 m as of midday due to the anticipated solar radiation. The main danger is the increasing water seepage into the snowpack and the massive loss of firmness which is the result. Naturally triggering slab avalanches can be expected, particularly on very steep, sunny slopes below about 2700 m. Special caution is urged since the impulse of a wet-snow avalanche can sweep along the entire wet snowpack and the avalanche can then grow to large size. Caution is urged not only for ascents to refuges, but also for exposed transportation routes. More favorable conditions prevail only above 2800 m today."

Both avalanche accidents are large-sized, naturally triggered, wet-snow slab avalanches. It seems likely (but is not necessarily the cause) that one impulse of a loose wet-snow avalanche led to a slab avalanche fracturing in a deep persistent weak layer in the rocky summit zone of the Gargglerin, which then caused the accident.

Terrain. The avalanche fractured on an extremely steep, east-facing slope at about 2300 m altitude around midday. This is terrain which at that time was being subjected to maximum energy input. ▌

At the avalanche fracture in extremely steep terrain. The snowpack is thoroughly wet.

background knowledge: springtime situation.

dp.10 definition

The term "typical springtime situation" generally refers to the period when winter is slowly coming to an end and rising temperatures, along with increasingly intense solar radiation, bring about quite unique conditions in backcountry touring regions.

Variety of forms

A springtime situation has at least two faces which need to be examined separately. The first face is the increasingly intense solar radiation which sets in about mid-February. The warmth which results leads to heightened creeping movements inside the snowpack, thereby increasing the likelihood of triggering by skiers and boarders. Naturally triggered avalanches are the exception, however. Primarily hit are pieces of terrain where the snowpack layering in winter was especially poor. Typically, it is very steep terrain on north-facing slopes below 2200 m in inneralpine regions, i.e. regions where snow is shallower. In winters where snowfall has been heavy, and when a pronounced weak layer has great masses of snow deposited on top of it, the situation is correspondingly delayed.
The problem zones then appear at higher altitudes as the process unfolds.

The second face of springtime situations is revealed regularly in very active periods of avalanche activity, usually starting in mid-March. Quite often, this is when the major part of all the season's naturally triggered avalanches occur in the form of wet-snow releases. Nowadays such situations can be recognized quite easily with the help of weather station graphs. Decisive are the high air temperature; the increasing air humidity; the high intensity of solar radiation; and low wind impact. During this process the snowpack surface temperature nears and often reaches zero degrees. That is the moment when naturally triggered avalanches are just waiting to happen. If the snowpack is

The start of naturally triggered avalanche activity is often a snowpack surface temperature of 0°C.

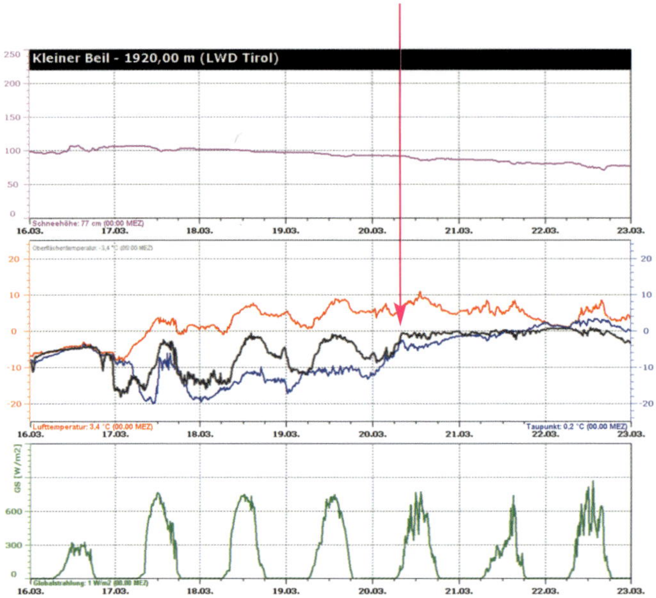

loosely structured, the avalanches will be wet, loose-snow releases. If the snowpack is bonded and there is at least one weak layer inside it, wet slab avalanches are also generated.

The spring situation is determined by four major meteorological parameters, quite removed from heavy snowfall: temperature, moisture, solar radiation and, in some cases, wind.

☐ Temperature

Air temperature measurement is defined quite precisely in meteorology: the temperature in the shade (this used to mean inside a weather hut) two meters above the ground. In modern automated weather stations no weather hut is used, since they offer too large a surface area for often stormy high altitude winds. But even without a weather hut, protections against solar radiation, sometimes including ventilation, ensure that the temperature which is measured is "in the shade."

Air temperature ordinarily recedes with ascending altitude. If the air is dry, it decreases by approximately 1°C/100 m; if the air is humid, by 0.65°C/100 m (moist adiabatic temperature gradient). So-called inversion weather is an exception, particularly in winter.

Automated weather stations supply important data for assessing avalanche danger.

Humidity

It is important to distinguish between absolute and relative humidity. Absolute humidity states how many grams of water vapor are contained in one cubic meter of air. This value depends on temperature: the warmer the air, the more moisture it can absorb. Thus, about 1 m^3 of air can hold maximum 4.8 grams of water vapor at 0°C; at 30°C it can hold 30.3 grams! This is the highest possible moisture content. Relative humidity is used more frequently than absolute humidity: it states in percent how much of highest possible water vapor is actually contained in the air. In other words

$$\text{relative humidity} = \frac{\text{absolute humidity}}{\text{highest possible humidity}} \cdot 100 \text{ (in \%)}$$

Dew point

The dew point is the temperature to which the air must cool for saturation, i.e. 100% relative humidity, to be reached. When air is saturated, air temperature and dew point are identical. The drier the air, the further the dew point is below air temperature. If the dew point is reached or undershot, condensation (cloud formation) takes place.

Radiation

A distinction is made between solar radiation (short-wave radiation) which is energy input into the earth; and terrestrial radiation (long-wave radiation=outgoing warmth radiation, see dp.8) in which the earth loses energy. The difference between these two radiation variants is known in meteorology as "radiation balance."

The radiation balance (not solar radiation!) is what is decisive for the warmth of the earth's surface and thereby, also for the temperature of the lowermost layer of air in the atmosphere. If the radiation balance is positive, the earth's surface absorbs energy and warms up. Conversely, when the earth releases energy it cools down.

In general, air temperature and solar radiation are frequently thrown into the same pot. But one needs to keep the two values quite strictly separated. As we all know, it can also be bitter cold when sunshine is intense. At over 3000 m in May, brilliant sunshine and temperatures below 0°C are quite normal. But whoever forgets to apply sunscreen will soon make a very unhappy acquaintance with solar radiation through powerful sunburn.

Avalanches as a result of radiation in springtime including radiation balance of a flat snowpack in Tirol at midday in February.

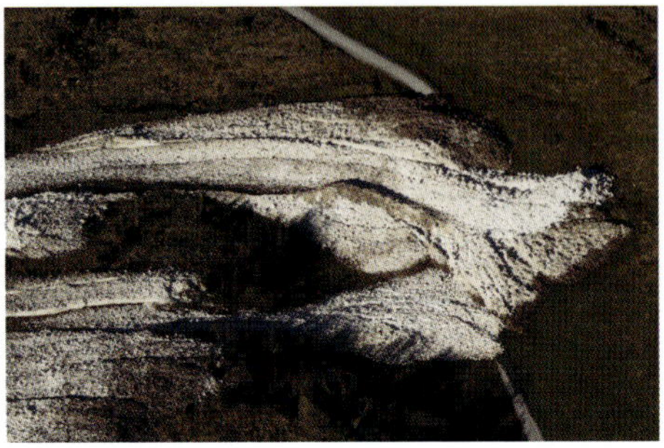

Radiation intensity

Radiation reaching the earth from the sun is reduced along its path through the atmosphere by gases in the air, by water vapor, by clouds and by dust. The proportion of radiation which actually reaches the earth's ground or the snowpack's surface depends on altitude, slope angle and, of course, season and time of day.

Whereas in deep winter, for example, less than 300 W/m² of short-wave solar radiation is measured, this measurement can increase to 1000 W/m² in springtime. Another statistic: in high winter, a 30° south-facing slope gradient receives nearly two and a half times the radiation as a flat surface. What is also important for the energy input from solar radiation is the measure of reflection, so-called "albedo": whereas fresh, dry snow reflects about 90%-95% of solar radiation, wet old snow reflects only about 50%-80%.

The remaining radiation hardly penetrates deeper than 30 cm into the snowpack, depending on the size of the snow crystals (the smaller the crystals, the less deep radiation penetrates). Most of the radiation is absorbed only in the uppermost centimeters of the snowpack. Whereas in the early morning hours following a night of clear skies, safe conditions prevail on a superficially melt-freeze encrusted snowpack, only a few hours later the avalanche situation is absolutely critical.

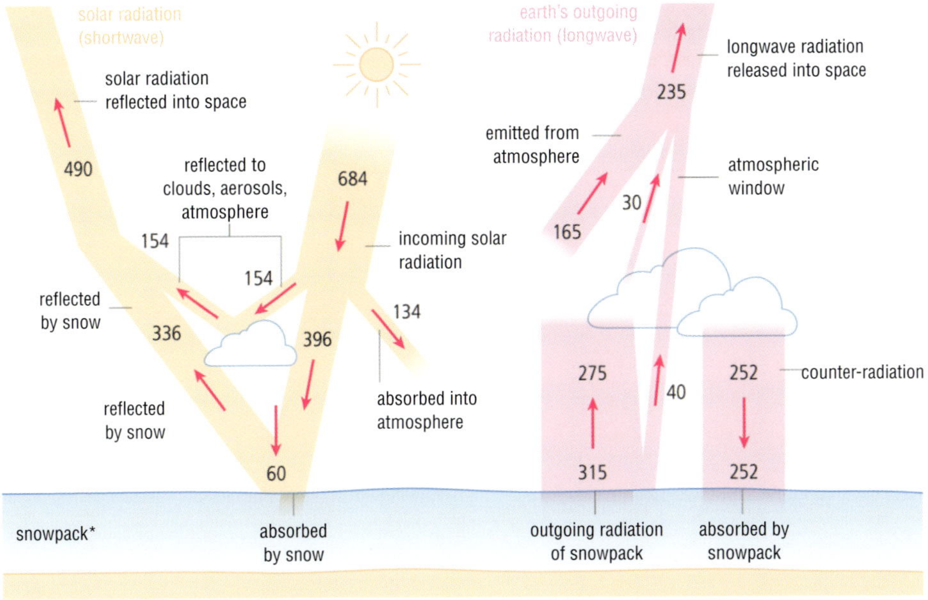

* 0°C | Tirol | February | slightly cloudy skies

all values in W/m2 at approximately midday

Beginning on east-facing and southeast-facing slopes, the day unfolds with rapidly rising air temperatures combined with increasingly intense solar radiation, both of which soften the melt-freeze crusts capable of bearing loads, the snowpack forfeits its overall firmness.

This process unfolds even faster if, in addition, moisture in the air is high. It causes the snowpack to become even wetter, and no condensation takes place. Suddenly, the activity of naturally triggered avalanches is accelerated. If the snowpack is thoroughly wet all the way down to the ground (when the snow temperature is 0°C in all layers), the avalanches can fracture all the way down to the most deeply embedded layers and grow to large size (ground avalanches).

This situation of a superficially moistened or thoroughly wet snowpack is the only one where wind can actually have a positive influence. If winds waft over the snowpack surface, this increases condensation, which cools the snowpack and consolidates it slightly. Ordinarily, a shiny, very thin layer of ice forms, a so-called firn-mirror. The drier the air, the greater the condensation.

Since condensation requires about eight times as much energy as melting, the snowpack has a great deal of energy (warmth) extracted from it through condensation. It cools and consolidates. That is why corn-snow conditions are better after a night of clear skies and dry air and also persist for longer than when the air is high in moisture. ▌

avalanche griesskopf.

Avalanche. On 12 March a group of eight backcountry skiers from Germany climbs from Kaisers in the Ausserfern region to the Griesskopf. One part of the group remains at the Kaiserjochhaus below the summit slope, the others reach the summit just before noon. When the group leader skis onto the slope at about 12:30 pm as last member of the group, he triggers a slab avalanche. He is swept along and buried. The other members of the group are outside the danger zone when the avalanche lets loose. In spite of swift action, the buried skier does not survive.

Short analysis. This avalanche is the opening note unleashing a phase of heightened likelihood of slab avalanche triggerings because the snowpack is starting to become thoroughly wet. In Tirol´s northernmost regions where the snow is shallow and where the accident slope is located, this applies to very steep west-facing and east-facing slopes at 2300 to 2700 m to begin with. On the day of the accident the day is cloudless, following a night of clear skies. The accident occurs at about 12:30 pm, when perfect corn snow blankets the slope. Nevertheless, the solar radiation at this time of day is already quite intensive, the air temperature is high. These two combined factors are sufficient to make the snowpack surface thoroughly wet on extremely steep slopes. What matters most for this particular accident is the small amount of snow lying on top of a deep persistent weak layer in this aspect. When water seeps down to this layer, it becomes immediately prone to triggering after a long period of stability. ▌

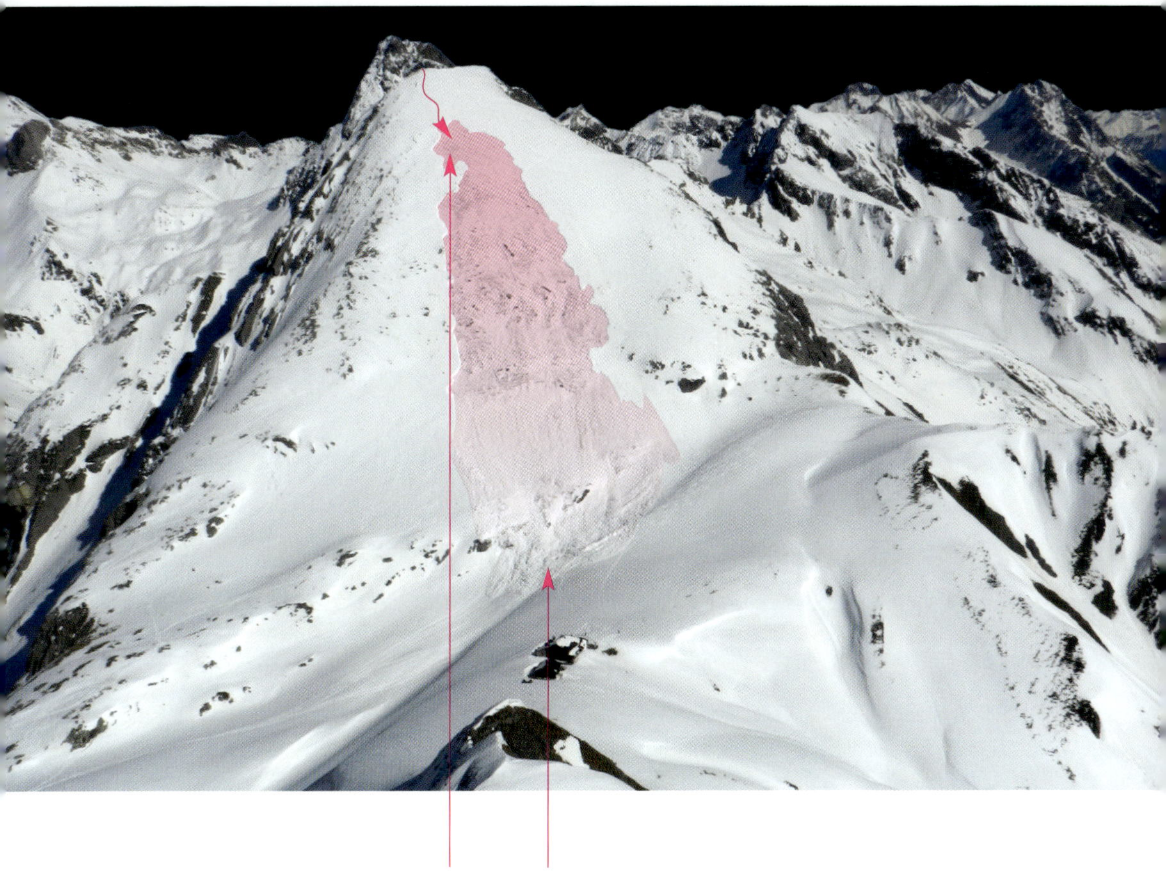

Where Griesskopf / Ausserfern / 2520 m / W-slope / 45°

Who 8 persons involved / 1 fatality

When 12. 3. 2014, 12:30 p.m.

Avalanche slab avalanche (wet) / L 300 m / W 30 m / fracture 0.2–0.4 m / burial 1 m / 15 min.

Regional danger level 2 (moderate)

Aval. bulletin headline Springtime situation: favorable in the morning, heed daytime rise in danger!

Avalanche problem wet snow / persistent weak layers

avalanche elmer muttekopf.

Avalanche. Eleven backcountry skiers climb in the Lechtal Alps over the Schafkar to the Elmer Muttekopf. The last section of the tour leads over a steep slope to the summit at the foot of a crest which is heavily dotted with cornices. The group has dispersed somewhat along the way, some of them are on the summit slope, others follow them at irregular distances. Suddenly a cornice breaks above the ascent route. Through the additional loading it adds to the snow cover, a wet-snow avalanche is triggered. The others immediately go into action, send an alarm signal and begin the search. Even though they discover and free the skier swiftly, they lose the race against time, the skier dies.

Short analysis. Since the middle of March, ongoing warmth has made the snowpack thoroughly wet to an increasing degree. This also applies to the very steep east-facing slope where the accident takes place, where the snowpack is moist or wet down to the ground. The days prior to the accident were instable, with rainfall extending up to altitudes of 1700 m. The previous day, on the other hand, had high pressure weather conditions, the snowpack thus cooled off sufficiently the night before. The group starts out early, moves across a melt-freeze crust which is capable of bearing their load. Cornices are an obvious sign of danger. Precisely when cornices will break, however, is utterly unpredictable, much like glide-snow avalanches. For that reason, the accident can also be ascribed to bad luck, for it is the breaking cornice which triggers the avalanche. ▌

spot of burial

Where Elmer Muttekopf / Arlberg – Ausserfern / 2220 m / E-Hang / 40°

Who 11 persons involved / 1 fatality

When 1. 4. 2005, 9:00 am

Avalanche slab (wet) / L 200 m / W 30 m / fracture 2 m / burial 1 m / 15 min.

Regional danger level 2 (moderate)

Aval. bulletin headline Increasingly frequent loose-snow avalanches

Avalanche problem wet snow

happy end

by the skin of our teeth

To assess avalanche danger is a complex task. That is just as true for professionals as for amateur winter sports lovers. Thanks to far-reaching information networks, the avalanche warning services can now depict each respective situation with a high degree of accuracy, providing decision makers and backcountry tourers with immense support. However, a 100% depiction of nature is impossible and can never be realistically expected. Thus, as long as there is snow in the mountains there will be avalanches and accidents, in spite of all the risk management we can devise. There remains a certain indissoluble risk and the necessity of dealing with uncertainty. For some, this is an essential spice in the undertaking, for others an undesirable, unavoidable aspect. The fact is, statistics demonstrate avalanche accidents to be decreasing, despite increasing numbers of people venturing into outlying terrain. The reasons for this are complex. Avalanche warning services deserve some of the credit, as does heightened awareness and training of winter sports lovers, along with ever better safety measures in ski areas and more heavily tracked terrain outside ski areas. All in all, accidents are few. Sometimes due to a lucky break. However, in order not to push one's luck too far, it is best to be as well prepared as possible.

avalanche jamtalferner.

Avalanche. About 100 skiers took advantage of a tiny window of fine weather for a backcountry tour to the deep valley enclosure of the Jamtal in moderately steep terrain, where they heard a particularly loud settling noise of the snowpack. Right afterwards, four large slab avalanches triggered at 10-second intervals, some of them large. Two of them buried three of the skiers. Once again, luck and quick action by other members of the group permitted all of them to be rescued.

Short analysis. For three days prior to these avalanches, storm-strength winds prevailed with lots of fresh snow. Inside the snowpack were numerous weak layers, both at ground level and at the uppermost surface. What was decisive for these triggerings was a pronounced layer of graupel lurking inside the fresh snow. The stormy winds were equally significant, along with the markedly higher temperatures that morning. Both encouraged the formation of a wide-surface slab. Through the impulse of the numerous ascending skiers, a fracture occurred in the layer of graupel which then propagated further, possibly deep into flat terrain. This, in turn, caused an additional, much greater impulse on the wide-area weak layers near to the ground, which then also fractured. This fracture then propagated over a course of several kilometers.

The avalanches which were triggered resulted from a combination of different avalanche problems and different danger patterns. Preeminent among them were a new snow problem, a wind-drifted snow problem and a persistent weak layer problem and their corresponding danger patterns dp.1, dp.6 and dp.9.

Where Jamtalferner glacier / Silvretta / 2900 m / W-N-E-facing slopes / 40°

Who appx. 100 persons involved / 1 slightly injured person

When 8. 3. 2017, 10:00 am

Avalanche slab avalanche (dry) / L 600 m / W 300 m / fracture depth 0.5–3 m / burial 1 m / appx. 10 min.

Regional danger level 3 (considerable)

Aval. bulletin headline Tense avalanche situation: considerable danger widespread!

Avalanche problem persistent weak layers

Danger pattern dp.1, dp.6, dp.9

avalanche schafleitenalm.

Avalanche. A local backcountry skier meets two snowshoe hikers on the trail to the Rastkogel refuge near the upper reaches of the Schafleitenalm. Shortly afterward, the skier remotely triggers a small slab avalanche in nearly flat terrain. At that moment he is at the foot of a small, steep slope in the fall line to the Schafleitenalm. The avalanche catches the skier and sweeps him along. He comes to a halt right in front of the entrance to the refuge, completely buried in snow. The door cannot withstand the pressure of the snow masses, he is hurled into the hut where he can free himself from the snow masses. In the interim, the snowshoe hikers have organized help and start to search for him in the conical-shaped avalanche path. The skier can leave the refuge through the front door with only slight injuries. The helicopter arrives and takes him down to the valley.

Short analysis. Truly a curious tale, fortunately with a happy end. The avalanche is but one of many which are registered in this period. Three days prior, a two-week long period of cold comes to an end. Thereafter up to 50 cm of fresh snow arrives over far reaching areas, accompanied by strong winds. Over widespread areas, the snowpack is extremely prone to triggering. During our investigation of this event, we are ongoingly escorted by numerous glide cracks and settling noises as we make our way to the scene of the accident. The settling noises can be explained by the escape of air from the concavities inside the weak layer. It is also typical of such situations that avalanches are remotely triggered from flat terrain. Also typical are naturally triggered avalanches which occur even on hillsides which are not steep.

Where Schafleitenalm / Tux Alps / 1700 m / south-facing slope/ 35°

Who 1 person involved, injured

When 18.2.2012, 11:50 am

Avalanche slab avalanche (dry-snow) / L 40 m / W 30 m / fracture 0.4 m

Regional danger level 3 (considerable)

Aval. bulletin headline Treacherous avalanche situation, considerable danger widespread

Avalanche problem persistent weak layers

Printing information

6th edition 2018 © Verlagsanstalt Tyrolia
Graphic design. grafische auseinandersetzung, christine brandmaier, 6410 telfs
Translation. Jeffrey McCabe
Cover. Avalanche on Mutteberger Seespitze on 4.4.2006
Illustrations. Lisa Manneh, in accordance with materials from Rudi Mair and Patrick Nairz
Photos. Patrick Nairz, Rudi Mair, N. N. (pp. 4, 5), Thomas Ebert (p. 7), Alpine Police (pp. 55, 69, 70, 71, 85, 86, 99, 111, 165, 227), Josef Probst (p. 81), LWD Bayern (p. 183), BR Lienz (p. 95), Reinhold Oblak (pp. 189, 190, 191), Lukas Ruetz (p. 193), Robert Aichholzer (p. 225)
Print and binding. Gorenjiski Tisk, Slovenia
ISBN. 978-3-7022-3696-0
E-Mail. buchverlag@tyrolia.at **Internet.** www.tyrolia-verlag.at